Pipe Line
Corrosion
—and—
Cathodic
Protection

**THIRD
EDITION**

G | **P**
P | 🛡

Gulf Professional Publishing
An Imprint of Elsevier

Pipe Line Corrosion and Cathodic Protection

A practical manual for corrosion engineers, technicians, and field personnel

THIRD EDITION

Marshall E. Parker
Edward G. Peattie

Gulf Professional Publishing and Butterworth-Heinemann are imprints of
Elsevier

Copyright © 1999 by Elsevier

Originally published by Gulf Publishing Company, Houston, TX.

 This book is printed on acid-free paper.

Library of Congress Cataloging-in-Publication Data
Parker, Marshall E.
    Pipe line corrosion and cathodic protection.
    Includes index.
    ISBN-13: 978-0-87201-149-6   ISBN-10: 0-87201-149-6
    1.   Pipe lines—Corrosion 2.   Pipe lines—Cathodic protection.
I. Peattie, Edward G.   II. Title
TJ930.P33   1984   621.8'672   83-22630

The publisher offers special discounts on bulk orders of this book.
For information, please contact:
Manager of Special Sales
Elsevier
200 Wheeler Road
Burlington, MA 01803
Tel: 781-313-4700
Fax: 781-313-4802

ISBN-13: 978-0-87201-149-6
ISBN-10: 0-87201-149-6

For information on all Gulf publications available, contact our World Wide
Web homepage at http://www.bh.com/gulf

Transferred to Digital Printing, 2010

Printed and bound in the United Kingdom

Contents

v

Preface to Third Edition

A pipe line buried in the earth represents a challenge. It is made of steel—a strong, but chemically unstable, material—and is placed in an environment which is nonuniform, nonprotective, and nonyielding. It is the duty of the corrosion engineer to study the properties of this system to ensure that the pipe line will not deteriorate.

In 1955, when I was first working on cathodic protection for pipe lines in Saudi Arabia, the first edition of this book by Marshall Parker was just a year old. Fortunately, the company library contained Mr. Parker's book. I found its simplicity and directness preferable in approaching a complex subject.

During the 30 years since its publication, generations of pipe line engineers and technicians have used this book as their first exposure to corrosion control. Many books on the subject have been published since 1954, but the Parker book is still the best introduction to the fundamentals. New technology has been developed, yet the principles of cathodic protection are still the same. The result is that we have more sophisticated instruments to use, but the measurements have not changed. Consequently, I have retained the still-valid material of the original Marshall text and made changes only when better and shorter methods are available.

The first task of the pipe line corrosion engineer is to study the properties of the earthen environment. First, we shall learn how to measure the resistivity of the soil as a preparation for further work. In the second chapter we shall study another electrical measurement: potential difference. To do this, we shall learn about the standard electrode for this measurement. In Chapter 3 we go on to line currents and extend these measurements to current requirement surveys in Chapter 4. Thus, the current necessary to design a cathodic protection system will be calculated.

Chapters 5 and 6 show the computations required for the design of an impressed current cathodic protection system. Then, Chapter 7 shows the design procedure for a sacrificial cathodic protection system.

The problem of partial cathodic protection of a pipe line by concentrating on "hot spots" is discussed in Chapter 8. A problem in Chapter 9, which does not usually occur nowadays, is that of stray-current corrosion; however, a corrosion engineer should still be able to identify this phenomenon and find its source. Interference (Chapter 10) is a problem corrosion engineers face every day and are still learning about.

The last two chapters of the book (Chapters 11 and 12) show the operation and maintenance of a cathodic protection system, and how to evaluate the coatings system in place. The Appendixes contain material basic to the knowledge of all corrosion engineers.

At the end of this book, the reader will be well on the way to being a capable corrosion engineer.

Edward G. Peattie
Professor of Petroleum Engineering
Mississippi State University

Preface to First Edition

How do people become pipe line corrosion engineers? Not by obtaining a degree in the subject, for no such degree is offered. Many corrosion engineers hold degrees—in chemistry, chemical engineering, electrical engineering, or any one of several others. Many others either have no degree, or have studied in some field apparently or actually remote from corrosion. All of these men became corrosion engineers by on-the-job experience and by individual study—some by trial-and-error methods, having been assigned the responsibility of protecting some structure from corrosion; some by working with people already experienced.

That this should be the case is not surprising, particularly when the protection of underground structures against corrosive attack is still as much an art as it is a science. Say, rather, that it is a technology; most of the design procedures used are either empirical, or, at best, are based on empirically modified theory. Almost every cathodic protection system installed has to be adjusted, by trial, to do its job properly. It is common experience that no two jobs are alike; every new project contains some surprises, some conditions not previously encountered.

These things are true for two main reasons. First, we do not as yet know enough about the subject; there is still room for the development of more powerful analytical methods. Second, the soil is a bewilderingly complex environment, and structures placed therein affect one another in very complicated ways. Almost never can we measure directly a single quantity which we seek; what our meters usually indicate is the result of what we are looking for and a number of partially unknown disturbing factors.

The corrosion engineer, then, must be to a large extent self-trained. And he can never expect to complete that training. This manual is an attempt to present some workable methods of doing some of the things the pipe line corrosion engineer is called upon to do. In no case is the method given necessarily *the* method; it is merely *one* method. If, by use of this manual, some corrosion engineers avoid some of the pitfalls into which I have fallen, and save time and effort in obtaining and interpreting field data, then I will be happy that my time has been well spent setting down some of the experience—a small part of which has been his own, a very large share of which is that of many others—and thus providing, in a way, a shoulder to look over.

Marshall E. Parker
Houston, Texas
January 18, 1954

1

Soil Resistivity Surveys

Pipe lines may be cathodically protected in either earth or water. To determine if cathodic protection may be used to prevent corrosion of steel pipe lines, we must first learn how to measure the resistivity of these environments.

Soil Resistivity Units

The unit of soil resistivity is the *ohm-centimeter*, usually abbreviated *ohm-cm*; the resistivity of a given soil is numerically equal to the resistance of a cube of the soil one centimeter in dimensions, as measured from opposite faces (Figure 1-1). The resistance of a rectangular solid other than a cube is given by

$$R = \frac{\rho \times L}{W \times D} \tag{1-1}$$

where W, L, and D are the dimensions in centimeters, as shown in Figure 1-1, and ρ is the resistivity; the unit must be ohm-cm in order for the equation to be consistent. The resistance between any two terminals, of any size and shape, in contact with a body of soil, of any extent, is determined by the size and spacing relationships and by the resistivity of the soil. For simple cases, the resistance can be computed, but the mathematical complexities are often very great.

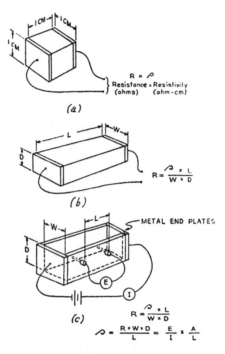

Figure 1-1. (a) Resistivity (ϱ) in ohm-cm is numerically equal to the resistance (R) in ohms between opposite faces of a cube one cm on the side. (b) Resistance of a rectangular solid. (c) Soil box, in which ϱ is obtained by measuring the resistance between the planes of the two potential pins.

Two-Terminal Resistivity Determination

An apparatus similar to that illustrated in Figure 1-1 may be used to determine the resistivity of soil samples; in corrosion investigations, however, it is much more useful to measure the soil in place.

If two terminals are placed in the soil, then it will be possible to determine its resistivity by measuring the resistance between the two electrodes. If they are placed close together, it will be necessary to maintain a known and fixed distance between them; if they are far enough apart, surprisingly enough, the indicated value is virtually independent of the distance. Either AC or DC may be used, and the resistance may be de-

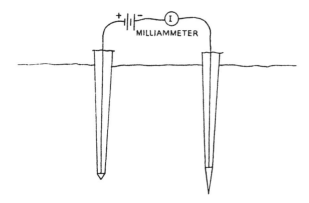

Figure 1-2. Shepard canes for soil resistivity measurement. Current from a three-volt battery (two flashlight cells) is passed through the soil between two iron electrodes mounted on insulating rods. Current flow is measured by a double range milliammeter (0–25 and 0–100) graduated to read directly in ohm-cm (10,000–400 and 500–100). Cathode is made larger to avoid polarization; accuracy is about 6% when tips are separated by 8 inches or more. Meter, batteries, and switch are mounted on anode rod.

termined by measuring current and potential, or by a bridge circuit, in which the unknown resistance is compared to one which is within the instrument.

Figure 1-2 illustrates Shepard canes, which use DC impressed between two iron electrodes. This apparatus is rapid in use, but gives the value corresponding to a small sample of soil, in the immediate neighborhood of the electrodes. Figure 1-3 shows an AC bridge-type apparatus; this device is also very rapid in use, but it, too, measures only a small sample. Numerous variations of the latter are on the market.

Four-Terminal Resistivity Determination

To include a larger sample of the soil and measure the resistivity at greater depths, the four-terminal or Wenner method may be used. The basic circuit is shown in Figure 1-4; the following figures (Figures 1-5, 1-6, and 1-7) show some of the specific methods based on this technique. Details of operation vary with the particular instrument used, but

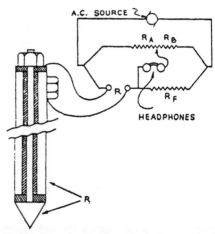

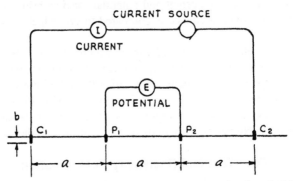

Figure 1-3. AC soil rod. Current from an AC source (usually a battery, buzzer, and condenser) is passed through the soil between a steel rod and an insulated steel tip. The slidewire is then adjusted until no signal is heard in the headphones; at the point of balance

$$\frac{R}{R_F} = \frac{R_A}{R_B}, \text{ or } R = R_F\frac{R_A}{R_B}$$

In practice, the slidewire $R_A R_B$ is graduated to read R directly; then ϱ = $R - C$, where C is a constant for the particular rod, determined by calibration in a solution of known resistivity.

Figure 1-4. Four-terminal (Wenner) measurement of soil resistivity. Distance (b), depth of electrode, must be small compared to (a). The basic formula is

$$\varrho = 2\pi a\frac{E}{I}$$

The resistivity is "averaged" to a depth approximately equal to the electrode spacing (a).

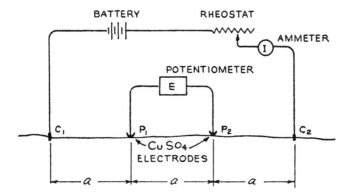

Figure 1-5. Ammeter-voltmeter measurement of resistivity. A storage battery of dry cells may be used for current. Copper sulfate electrodes will help avoid polarization errors. The voltmeter used must be high resistance, or a potentiometer may be employed. If the distance (a) is made 5 feet, 2½ inches, then the formula simplifies to

$$\varrho = 1000\frac{E}{I}$$

Direction of current should be reversed, and readings averaged, to balance out extraneous currents or potentials in the soil.

the principle is common to all. It should be noted that the resistance measured is that between the two inner, or potential, electrodes; the outer pair serve to introduce the current into the soil.

The value obtained is an ''average'' value to a depth approximately equal to the electrode spacing; actually, it is influenced to some degree by soil lying at even greater depths. There is no sharp dividing line, but the influence of soil at greater and greater depths becomes smaller and smaller.

Other Methods

There are at least three possible indirect methods of measuring soil resistivity in which no contact is made with the soil at all. An induction-type pipe locater gives an indication of the resistivity below it and may be calibrated to yield fairly accurate results. In general, this method may be very useful in scouting out low-resistivity areas, which can then

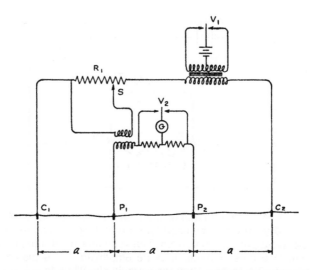

Figure 1-6. Vibroground (simplified diagram). Vibrators V_1 and V_2 are synchronized; V_1 converts the battery's DC into AC; the slidewire S is adjusted until the IR drop across R_1 just bucks out the current in the potential circuit, as indicated by a zero deflection on the galvanometer G; R is then read directly on the slidewire, and ϱ is obtained from $\varrho = 2\pi a R$.

be explored more precisely by one of the methods already described.

The value and phase angle of the impedance (high-frequency) between two parallel conductors lying on or parallel to the surface is a function of the soil resistivity, and thus a device could be built which would employ this principle. It is also true that the tilt of the radiation field of a remote radio transmitter at the surface of the earth is a function of the soil resistivity. Both of these methods offer some promise of development into very rapid and possibly automatic systems for the recording of resistivity, but the development has not yet been done.

Locating "Hot Spots" on Bare Lines

Bare lines, particularly gathering lines, are often so situated that the point of maximum economic return for cathodic protection is not the prevention of all leaks, but the prevention of 85–95% of them. This can often be done at a cost as low as 15% of that of full protection by the application of galvanic anodes—usually magnesium—to the "hot spots" or areas of maximum corrosiveness. This has been shown to correlate highly with resistivity so that a system of locating areas of low, and of relatively low, resistivity, is the indicated technique. Such a procedure is known as a soil resistivity survey, and consists of a series of measurements taken along the line (or right-of-way for a planned line), using any of the methods described. The four-point method is most commonly used.

Readings should be taken according to a systematic procedure; there is some conflict here between the demands of science and those of practicality. If readings are taken at uniform spacing, the statistical analysis of the entire set will be more representative of the corrosive conditions along this line, as compared to other lines. In other words, such a procedure will give the maximum of general information. The maximum of *useful* information, however useful in the design of a hot spot protective

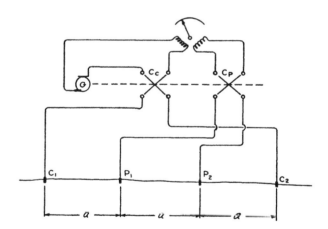

Figure 1-7. Soil resistivity Megger (simplified diagram). DC from the hand-cranked generator G is converted to AC by the commutator C_c, and enters the soil through C_1 and C_2, after passing through the current coil; the AC potential picked up between P_1 and P_2 is converted to DC by commutator C_p and is fed to the potential coil. The needle comes to rest, under the opposing forces from the two coils, at a point which indicates the resistance R. ϱ is then obtained from $\varrho = 2\pi aR$.

system, is not obtained in this way, but rather by some such method as the following:

1. A reading to be taken at least every 400 feet.
2. A reading to be taken wherever there is a visible change in soil characteristics.
3. Two successive readings should not differ by a ratio greater than 2:1 (when a reading differs from the preceding by more than this, then it is necessary to go back and insert a reading; this should be repeated until condition 3 is met).
4. As an exception to the last rule, no two readings need be taken closer than 25 feet.
5. As another exception to the 2:1 rule, it does not apply when the lower of the two readings is greater than 20,000 ohm-cm. The figures given in this set of criteria may be varied for specific cases; the foregoing set has been found useful and economical in the survey of over a million feet of gathering lines.

For this type of survey, readings would be taken which correspond to the soil at pipe depth. This means that holes will have to be drilled or punched for the single-rod apparatus. The four-terminal system should be used with a spacing of about $1\frac{1}{2}$ times pipe depth; very often a 5-foot, $2\frac{1}{2}$-inch spacing (which makes the multiplier $2\pi a$ just equal to 1000) is well suited for this purpose.

Data obtained in the survey are plotted on a diagram representing the length of the line, as illustrated in Figure 1-8. The vertical resistivity scale is logarithmic, since resistivity ratios are of interest, rather than their differences. From such a diagram the "hot spots" can readily be located. This technique will be discussed more fully in a subsequent chapter.

Surveys for Ground Beds

The resistivity of the soil in which a ground bed is to be installed is one of the primary design quantities involved. It is most important to get as low a total resistance as possible in order to obtain the necessary current output with the minimum amount of power. Unlike the survey to determine corrosiveness, as in the preceding topic, in this case the

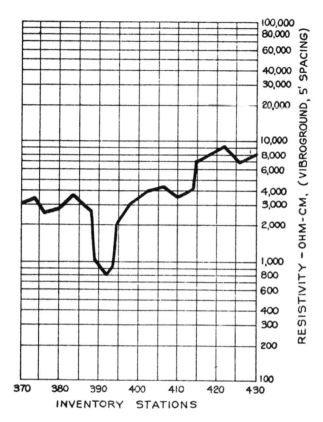

Figure 1-8. Soil resistivity profile. Measurements taken by any of the methods described are plotted as ordinates (using a logarithmic scale), with distances along the pipe line as abscissas.

resistivity to great depth is important. For this reason, the four-terminal method is the only one useful here. It is customary to take readings at 5-, 10-, and 20-foot pin spacing (actual values used are 5 feet, 2½ inches; 10 feet, 5 inches; and 20 feet, 10 inches; giving multipliers of 1000, 2000, and 4000, respectively).

The prospective site should be marked off in a rectangular grid, with spacings of 50 or 100 feet, and readings taken at each grid intersection; where differences are too great, intermediate readings should be taken.

These diagrams should always be plotted in the field so that a complete picture will be obtained. "Contours" can be drawn, as indicated in Figure 1-9. If colors are used, the readings at all three pin spacings can be put on the same diagram; only one is shown here. Note that the values chosen for contours are logarithmic, as were the vertical scales in the line resistivity survey.

Area Surveys

Where a set of piping and other underground structures is spread out over an area, as in a refinery or chemical plant, a survey similar to the one just described will furnish information useful both in the prediction of corrosiveness and in the design of protective systems. The spacing used between readings is usually greater than for ground bed design. A

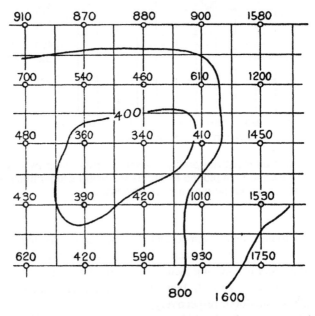

Figure 1-9. Resistivity plat. Resistivity readings taken at spaced intervals, with additional readings at critical points, are indicated by the small figures. The curves represent lines of equal resistivity. Such plats are of value in locating sites for ground beds and in their design.

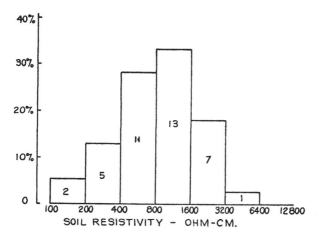

Figure 1-10. Resistivity histogram. Equally spaced resistivity readings taken over an area or along a pipe line route are plotted according to the frequency of occurrence of readings in the different ranges. Note that successive ranges cover equal resistivity ratios, rather than arithmetic differences.

preliminary spacing of 400 feet is often used, with intermediate positions used where necessary to get a clear picture. As in the ground bed survey, readings are usually displayed on a resistivity contour map of the plant site. Usually only one pin spacing is used, and the most frequent value is the nominal five-foot spacing already mentioned.

In addition to the geographical distribution shown on the resistivity plat, it is useful to prepare a histogram, or frequency distribution, as indicated in Figure 1-10. This shows the number of readings found in various "classes" or ranges of resistivity; these ranges should be logarithmic, as are the contour intervals on the plat already described.

Logarithmic Resistivity Ranges

Three examples have already been given of logarithmic treatment of resistivities. There is sound scientific basis for this approach, aside from the intuitive observation that ratios are more important than differences. It has been found that large numbers of resistivity readings, when expressed as logarithms, tend to fall into "standard" distributions. The standard distribution is a well-known and extremely useful concept in statistical analysis.

A consequence of this is that two extensive structures (such as pipe lines) each exposed to a wide variety of soils can be compared as to their corrosion exposure most accurately by comparing their *logarithmic mean resistivities*. The logarithmic mean resistivity of a set of resistivities is the value whose logarithm is the average of all the logarithms of the measured values. These can most easily be averaged by grouping them into classes, as shown in Figure 1-11, where the logarithmic mean of 115 different readings is found to be 14,900 ohm-cm.

It is almost universal practice to choose a ratio of 2:1 for the successive values used in any logarithmic system, for contours, histograms, or the calculation of a mean. Three different systems are in widespead use: the two already illustrated (A) based on 100 ohm-cm, as used in Figures 1-9 and 1-10; (B) based on 1000 ohm-cm, as used in Figure 1-11; and (C) which is not truly logarithmic, in that the ratio is not always the same. All three are shown here for comparison:

(A)	(B)	(C)
50	62.5	50
100	125	100
200	250	200
400	500	500
800	1,000	1,000
1,600	2,000	2,000
3,200	4,000	5,000
6,400	8,000	10,000

Scale (C) has the advantage in that all of the values are "round" numbers whereas if either (A) or (B) is extended far enough in either direction, the numbers tend to become awkward. On the other hand, when a mathematical analysis is made of the data, the irregular ratios involved in scale (C) introduce complications. Note, when (C) is used for a histogram, the horizontal widths of the "cells" should be made proportional to the logarithms of the ratios. Log 2 = 0.301, log 2.5 = 0.398; so the ratio of the two spaces should be about 3:4.

The corrosion technician now has various methods at his disposal to measure the resistivity of the soil in which the pipe line is to be in-

SUMMARY OF SOIL RESISTIVITY MEASUREMENTS

Calculation of Logarithmic Mean

Boundary Values Resistivity ohm-cm		N Number	L Log Mean	N x L Product
1,024,000 – 512,000			5.85	
512,000 – 256,000	//	2	5.55	11.10
256,000 – 128,000	////	4	5.25	21.00
128,000 – 64,000	////\ /	6	4.95	29.70
64,000 – 32,000	////\ ////\ ////\ /	16	4.65	74.40
32,000 – 16,000	////\ ////\ ////\ ////\ ///	23	4.35	100.05
16,000 – 8,000	////\ ////\ ////\ ////\ ////\	25	4.05	101.25
8,000 – 4,000	////\ ////\ ////\ ////\ ////\ //	27	3.75	101.25
4,000 – 2,000	////\ ////\ /	11	3.45	37.95
2,000 – 1,000	/	1	3.15	3.15
1,000 – 500			2.85	
500 – 250			2.55	
250 – 125			2.25	
125 – 62.5			1.95	
62.5 – 31.25			1.65	

Totals	115	479.85
Logarithm of Mean		4.172
Logarithmic mean resistivity, ohm-cm		14,900

Figure 1-11. Summary of soil resistivity measurements.

stalled. The question may now be asked: What if the pipe line is to be installed in water—would the same methods of measurement work? The answer is yes, but with reservations. Actually, a single sample of the water is enough, and it can be measured in a soilbox, as shown in Figure 1-12. The countless variations of resistivity which we have shown in Figure 1-9 for soil do not exist in the water environment, and so one resistivity value in ohm-centimeters is usually enough for the engineer to use. For soils, as we have seen, it is more complicated.

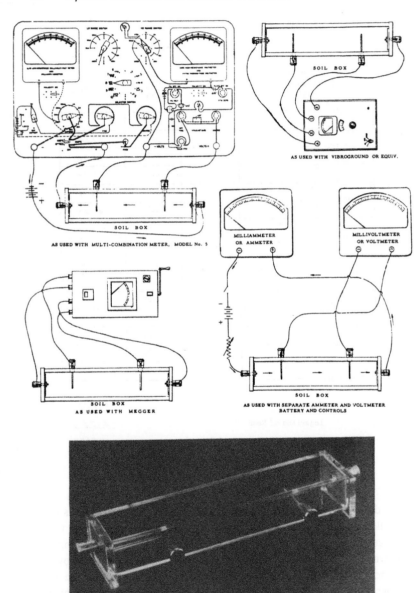

Figure 1-12. Typical connections for use of soil box with various types of instruments. (Courtesy of M. C. Miller Co., Inc.)

Summary

The resistivity of soil is usually measured by the four-pin method with the Vibroground (Figure 1-6), possibly the easiest instrument to use. It is possible to attach a set of pins and leads which will minimize the time to get soil resistivities, and a complete survey may be done in a relatively short time. The newest example of the soil resistance meter is shown in Figure 1-13 and has been available since 1981. Its advantage is that there are no moving parts and no vibrators to change. Although called a four-pin soil resistance meter, it is also excellent for measuring water samples in a soilbox.

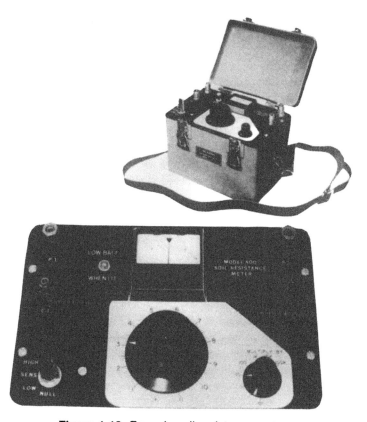

Figure 1-13. Four-pin soil resistance meter.

2

Potential Surveys

Pipe-to-Soil Potentials: Electrodes

The potential difference between a buried pipe and the soil is of considerable importance, either in investigating the corrosive conditions or in evaluating the extent of cathodic protection being applied. This quantity actually is measured by connecting an instrument between the pipe itself (direct metallic contact) and a special electrode placed in contact with the soil. This electrode is also called a *half cell*.

The most common type of electrode in use is that which employs a metal-to-electrolyte junction consisting of copper in contact with a saturated solution of copper sulfate; this particular combination is made of easily available materials and is very stable. Figure 2-1 shows two typical electrodes. Other types in use are:

1. The hydrogen electrode, used only in laboratory investigations.
2. The calomel electrode, used often in fresh water or saltwater.
3. The lead/lead chloride electrode, frequently employed in studying the corrosion of lead cable.
4. The silver/silver chloride electrode, used in seawater because it is not subject to contamination by salt incursion.
5. The pure zinc electrode (in packaged backfill). This is suitable as a permanently installed reference electrode. (Note: all these listed electrodes must be corrected to give readings similar to the copper/copper sulfate half cell. See Appendix C.)

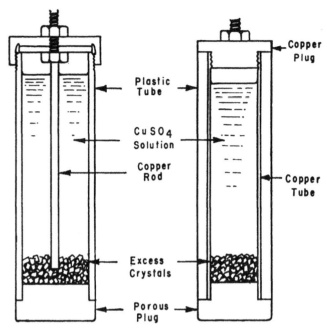

Figure 2-1. Copper sulfate electrodes. The essential parts are: (1) a piece of copper, to which the instrument is connected; (2) a saturated solution of copper sulfate, in contact with the copper; (3) a porous member, placed in contact with the soil. It is essential that no metal touch the copper sulfate solution except copper. Excess crystals are added, to ensure that the solution always will be saturated.

Voltmeters

One choice for the instrument to be used in measuring this potential is the voltmeter, as shown in Figure 2-2; there are two possible sources of error in its use. First, while a voltmeter correctly indicates the potential difference *across its terminals*, the current which flows through the instrument (a voltmeter is essentially a milliammeter) introduces an IR drop in the rest of the circuit, which is not included in the reading. For example, if the resistance of the external circuit (pipe, lead wires, soil, and electrode) is one-fourth the resistance of the meter, then the meter will read only four-fifths of the total potential; the error will be 20%.

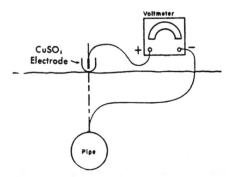

Figure 2-2. Voltmeter measurement of pipe-to-soil potential. Connection to the pipe may be by means of a welded or soldered lead wire (as illustrated), by a contact bar, or by mechanical connection to a valve or fitting above the surface. The latter should be used with caution, as flanged fittings sometimes introduce extraneous potentials. If the soil is dry, a little water may be added to lower the contact resistance; it is usually adequate to dig a little below the surface.

A second source of error is that the passage of current through the circuit may polarize the electrode or even the pipe itself, and so change the potential we are trying to measure. Both of these errors may be minimized by the use of a high-resistance voltmeter; not less than 50,000 ohms per volt is recommended, and this is often too low. Readings should always be taken on two different scales (introducing two different values of meter resistance); if they are in substantial agreement, the value may be accepted.

In the next section, methods will be given for obtaining the corrected value when the error due to voltmeter resistance is not too high (see Equations 3-2 and 3-3).

There is also available a special "dual sensitivity" voltmeter which offers two different resistance values on the same scale. If the reading is the same for these two (the change is made by merely pushing a button), then no correction is needed. If pushing the button makes a small change in the reading, then a very simple correction can be made. And, finally, if the change is large, an equation comparable to those cited previously can be used. This meter is very useful for making rapid potential readings.

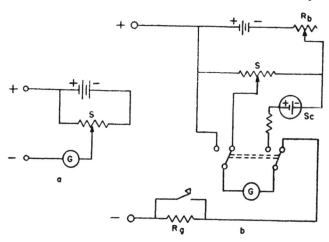

Figure 2-3. Slide-wire potentiometer. (a) Basic circuit. When the slide-wire S is adjusted so that the galvanometer G shows no deflection, then the unknown potential connected to the terminals can be read directly on the slide-wire scale. This simple circuit could be used only if the battery voltage were constant. (b) Modified circuit, with provision for adjusting battery voltage with resistor R_B, using standard cell SC for calibration. Resistor R_O protects the galvanometer against excess currents; it is shorted out by the switch when the instrument is almost perfectly balanced. Most instruments for corrosion use have two or more scales, instead of the one shown here.

Slide-Wire Potentiometer

These problems largely are overcome by the use of the potentiometer, an instrument which draws no current from the circuit at the point of balance. It should be remembered, however, that current is drawn while balancing is in progress and that some polarization may occur. Figure 2-3 shows this instrument and indicates its use.

Potentiometer-Voltmeter

Another approach to the problem is the potentiometer-voltmeter, one form of which is shown in Figure 2-4. Both of these useful instruments suffer a disadvantage when fluctuating potentials are encountered, as in a stray-current area, since it is impossible to follow rapid fluctuations with them.

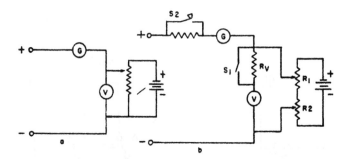

Figure 2-4. The potentiometer-voltmeter. (a) Shows the basic circuit. When the galvanometer deflection is zero, the external potential connected to the terminals is the same as that portion of the battery voltage which is indicated on voltmeter V; this, then, is a direct reading of the unknown potential. In (b) several refinements have been added: (1) with switch S_1 open, a high range is added; (2) the two resistors R_1 and R_2, of different values, provide a fine and coarse adjustment, and (3) switch S_2 controls a protective resistor, similar to that in Figure 2-3. More than two ranges are possible, and it is also possible to use the same instrument as an ammeter by building in shunts.

Vacuum-Tube Voltmeter

This instrument (not illustrated) has, through the use of transistors and other recent developments, now reached a stage of reliability and stability which makes it quite useful in the field. Battery-powered units are available with input impedances as high as 10 megohms, so the circuit resistance error ceases to be a problem. They possess an additional advantage over the various potentiometer types of instruments in that they do not have to be balanced, and hence may be used to follow fluctuating potentials without difficulty.

Solid-State Voltmeter

This instrument (Figure 2-5) is the successor to the vacuum-tube voltmeter and differs only in the substitution of solid-state transistors for the more vulnerable vacuum tubes. The resistance problems have been virtually eliminated, and accurate results may be obtained by nontechnical personnel. A particular model has selectable input resistances of 1, 10, 25, 50, 100, and 200 megohms. This allows the user to determine pipe-to-soil potentials in such difficult environments as city pavements. The range in reading varies from 0.1 mv to 20 mv. This is a rugged instrument for the technician who has to take many readings.

Figure 2-5. Solid-state DC voltmeter. (Courtesy of M. C. Miller Co., Inc.)

Multicombination Meter

This is the instrument most often used by corrosion engineers because of its versatility. It is a development of the potentiometer-voltmeter already described but has the following particular characteristics, as shown by the latest 1983 model in Figure 2-6. The *right meter* has a liquid crystal display with 5 ranges from −20 mv to 200 v and with selectable input resistances of 1–200 megohms. The *left meter* also has a liquid crystal display and serves both as a voltmeter and an ammeter. The four voltage ranges include −20 mv to 20 v, and the ammeter range is from −20 ma to 20 a. As an ohmmeter, it reads from −20 ohms to 2000 ohms. This instrument is still available with the conventional moving pointer dials, which are not as easy to read as are the liquid

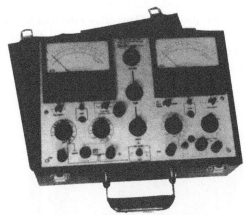

Figure 2-6. Multicombination meter. (Courtesy of M. C. Miller Co., Inc.)

crystal numbers. The instrument is used for pipe-to-soil potential differences, current versus IR drop, checking continuity of test leads and the resistance of bond wires, to name only a few examples.

Electronic Potential Meter

If the most advanced instrument for the corrosion engineer is the multicombination meter, the simplest of the new meters is the electronic potential meter, which is simply a small voltmeter attached to a copper-copper sulfate electrode. Figures 2-7 and 2-8 show it in two forms for determining potential difference in the field. It is rugged, inexpensive (about 12% of the cost of the multicombination meter), and ideal for inexperienced workers to use in average rural pipe line environments.

Electrode Placement

Figure 2-9 shows the current and potential fields surrounding a pipe line which are either collecting or discharging current at the point indicated. It will be seen that there is no point on the surface of the earth which is at the same potential as the surface of the pipe itself. Therefore, any electrode position chosen will inevitably introduce some IR drop error into the readings. This error is least (for a symmetrical field)

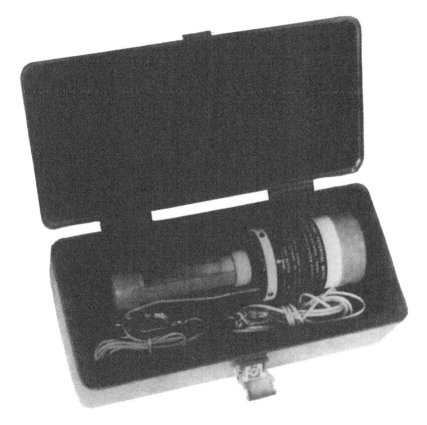

Figure 2-7. The electronic potential meter. (Courtesy of M. C. Miller Co., Inc.)

when the electrode is placed directly over the pipe. The only way in which the theoretically correct reading could be taken would be to place the electrode adjacent to the pipe, or to use a complicated "null circuit" involving two or more electrodes, in which the effect of the IR drop is canceled out. Both of these are quite time-consuming, and thus have gained little favor.

Figure 2-8. Potential meter with optional accessory electrode extension. (Courtesy of M. C. Miller Co., Inc.)

Pipe Line Connection

As shown in Figure 2-2, connection to the pipe must be made by an insulated lead wire or by an insulated contact bar or, best, by permanent test stations installed along the pipe line at regular intervals. The advantage of the test station is that, once it is installed, it will not be necessary to damage the pipe line coating to make contact with the line. Figure 2-10 shows a typical installation. It can be seen that a portion of the coating is scraped off, and the wire is then attached to the line by a process known as "cadwelding." This is a small thermite unit using

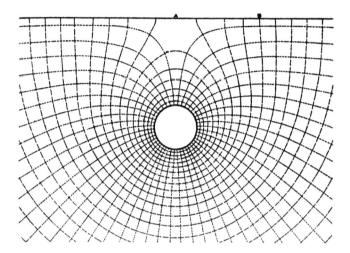

Figure 2-9. Current and potential field around a pipe. The dotted lines represent the flow of current to or from a pipe lying in soil of uniform resistivity; the dashed lines represent equal differences of potential. It can be seen that no point on the surface is at the same potential as that immediately adjacent to the pipe, but an electrode placed at A represents a closer approximation than one placed at B.

aluminum powder to produce enough heat to bond the copper wire to the steel. Then the connection and the surrounding bare metal are coated with a coating similar to the pipe line coating and usually finished with tape. The lead wire is then brought to the surface and is usually raised to a height of five feet, where a box containing a lead will make connection of the testing meter very convenient. When the test station is installed, it is usually convenient to use it as one end of a test section for measuring line current, as shown in the next chapter.

Surface Potential Survey for Corrosion

Figure 2-11 represents the current and potential fields around a section of pipe line with a single active corroding (anodic) area. The distribution of potential along the surface of the earth above the pipe clearly indicates the location of the active area; thus a survey of surface poten-

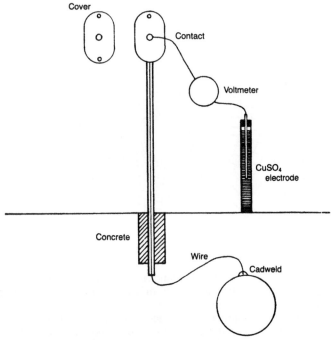

Figure 2-10. Test station installation.

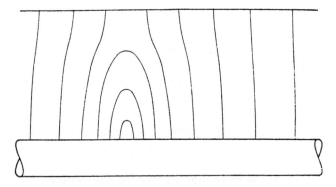

Figure 2-11. Potential field surrounding a single anodic point. An isolated anodic point on a buried pipe line can be detected easily by surface potential measurements, provided that readings are taken at sufficiently close spacing.

tials along the length of a pipe line should be of value in locating corrosion. Figure 2-12 diagrams the procedure for making such a survey, while Figure 2-13 shows a typical pipe-to-soil potential of a protected line.

Variations in soil resistivity introduce potential differences which may mask those sought, so the problem is by no means as simple as it might appear. Another drawback to this method—and a very serious one—is that it will not disclose a highly localized cell, where the anode and cathode are very close together, as for example, the common "concentration" cell where the anode is on the bottom of the pipe and the cathode on top, at the same location.

Pipe-to-Soil Potential as a Criterion of Cathodic Protection

It is almost universally accepted that a steel structure under cathodic protection is fully protected if the potential is at least 0.85-volt negative, referred to a standard copper-saturated copper sulfate electrode placed in the electrolyte immediately adjacent to the metal surface. The entire structure is fully protected, of course, only if this criterion is met at *every point on the surface.* This is not a minimum requirement; it is a maximum. In other words, it is almost certainly true that protection is complete when this is achieved. It is also possible to have complete freedom from active corrosion at lower potentials.

As mentioned, it is impracticably slow to place the electrode immediately adjacent to the pipe surface; it is completely impossible to take a reading at every point of the surface. As a consequence of these two difficulties, it is generally accepted that the electrode is to be placed in the soil immediately over the line; it is not difficult to show that, for a well-coated line, this is as good as next to the pipe. For a bare line, this is not true, but bare lines are rarely fully protected in any event. The other difficulty is avoided by assuming a certain degree of continuity in the potential gradient along a line. This assumption is almost always valid, and, it turns out, when it breaks down because of discontinuities in soil resistivity, it fails only in areas which are virtually noncorrosive without protection.

It is customary, then, and relatively safe, to consider a coated line to be fully protected if a survey along its length yields a reasonably smooth

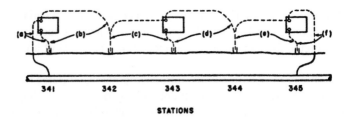

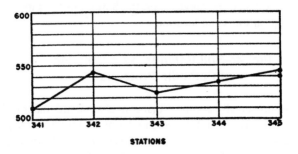

Procedure:

(a) Measure pipe-to-soil potential at test lead, Sta. 341.

(b) Measure soil-to-soil potential between 341 and 342; add (algebraically) to reading obtained at (a). This gives pipe-to-soil at Sta. 342.

(c) Move instrument two stations: measure soil-to-soil between 342 and 343.

(d) Measure soil-to-soil between 343 and 344.

(e) Move two stations: measure soil-to-soil between 344 and 345.

(f) Measure pipe-to-soil at 345. Reading should check derived value; error may be distributed, or, if small, ignored.

Notes may be kept as follows:

341	510 mv			
342		+ 35	545	
343		− 20	525	
344		+ 10	535	
345	540	+ 10	545	error = + 5 mv.

Two electrodes are used; they should be "leap-frogged" to avoid an error due to a small difference between the electrodes.

Figure 2-12. Surface potential survey.

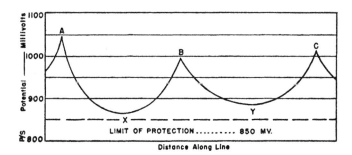

Figure 2-13. Pipe-to-soil potential of protected line. Typical distribution of potential along a line with cathodic protection units at A, B, and C. Points X and Y are the critical points; if the potential remains satisfactory at these points, the entire line is probably adequately protected.

potential curve which does not dip below the value of 0.85 volt; all readings being taken are with respect to a standard copper sulfate electrode.

Once a detailed survey over the system has established that the line is protected, then periodic checks at the critical points (low potential points between units) can make it possible to know that protection is still being afforded. This subject is discussed more fully later.

Other Applications of Pipe-to-Soil Potentials

This measurement forms a part of the technique used in several types of studies, most of which will be described later. Among them may be mentioned: (1) interference studies, (2) stray current investigations, (3) coating conductance measurements, and (4) cathodic protection tests.

Other Criteria

In addition to use of the −0.85-volt standard for adequacy of a cathodic protection system, two other methods are in use:

1. *Test coupon.* Burying corrosion test coupons along the pipe line. These should be weighed and connected to the line at locations possibly subject to extreme corrosion conditions. After a given exposure time, they should be removed and reweighed. Any loss

in weight may be used to calculate the degree of protection necessary.

2. *Potential change.* Another criterion (called "doubtful" by Uhlig[*] but often used in cathodic protection design) is to use a potential change of 0.3 volt when current is applied as a measure of cathodic protection. This involves an interrupted cathodic protection current measurement, as will be shown in Chapter 4. Also, this concept may be used in primary design calculations, as described in Chapter 5.

Summary

We have thus seen that, although there are several reference electrodes (half cells) available for potential difference measurements, only the copper/copper sulfate electrode is used for most of the readings the corrosion engineer will take on land. Also, for precision, the potentiometer, particularly in the form of a combination instrument, is preferable to the low-resistance voltmeters. The placement of the reference electrode is best made on the surface of the earth directly over the buried pipe line. Sometimes, the soil should be moistened, although preferably nothing should be done to affect the environment. For contact with the pipe line, it is advisable to have test stations installed at regular intervals along all major pipe lines. This will eliminate the problems of direct metal contact with the pipe itself, which will be necessary if the permanent leads are not available. Although the "leap-frog" system may be useful when there is a distance between test stations, it would be advantageous to have a reel of wire available to extend the measurements along the pipe line, particularly in regions near the midpoints between rectifiers. Modern methods of pipe-to-soil measurement use recording voltmeters, which may be used to complete a series of pipe-to-soil potentials along a pipe line in a very short time.

---

[*] Uhlig, H. H., *Corrosion and Corrosion Control: An Introduction to Corrosion Science and Engineering, 2nd Edition,* 1971, John Wiley and Sons, Inc., New York, p. 225.

3

Line Currents

Now that we know the most common pipe line measurement, pipe-to-soil potential difference, we shall study some other measurements necessary for corrosion control. The next is line current measurement.

Measurement of Line Current in Test Section

A test section consists of a length of uninterrupted pipe line (no joints other than welds, no fittings) between two points to which an instrument may be connected (see Figure 3-1). The connection points usually take the form of a pair of permanently attached lead wires; the use of contact bars is difficult and generally unsatisfactory. The resistance of the pipe itself (not the lead wires) between the two points must be known; it will usually be on the order of 0.001 ohm, or may be expressed in conductance units, as 1.00 ampere/millivolt. To obtain this value, the length of the section must vary according to the pipe weight; values between 50 and 400 feet are common.

Voltmeter Line Current Measurement

The difference in potential between the two leads, multiplied by the conductance of the section, will give the value of the current flowing in the line. This difference in potential may be determined with a mil-

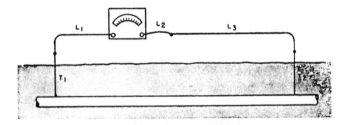

Figure 3-1. Measurement of IR drop in line section. R_w in Equation 3-1 is total resistance of T_1, T_2, L_1, L_2, and L_3, unless L_1 and L_2 are a pair of "calibrated" leads matched to and supplied with the instrument, in which case $R_w = T_1 + T_2 + L_3$ only.

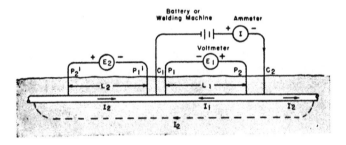

Figure 3-2. Calibration of test section. The conductance of the section, in amperes per millivolt, is given by

$$K = \frac{I}{E_1 + E_2} \text{ if } L_1 + L_2.$$

If these two sections are not equal, then

$$K = \frac{IL_2}{E_1L_2 + E_2L_1}$$

livoltmeter, as shown in Figure 3-2. For currents occurring naturally, a meter with a full-scale deflection as low as one millivolt may be required; the currents used in cathodic protection give rise to larger values. An instrument with a range such as this will have a low internal resistance, perhaps as low as one ohm; the resistance of the lead wires will thus be of considerable importance. If the value of the lead wire resistance is known, the corrected reading may be obtained from

$$V_{corr} = V_{ind} \times \frac{R_m + R_w}{R_m} \tag{3-1}$$

where:

V_{corr} = corrected reading
V_{ind} = indicated reading
R_m = meter resistance
R_w = total lead wire resistance

Another method of correcting for lead resistance, which has the advantage that the value need not be known, involves taking readings on two ranges of a multiscale voltmeter and applying

$$V_{corr} = \frac{V_L V_H (R - 1)}{R V_L - V_H} \tag{3-2}$$

where:

V_L = reading on low scale
V_H = reading on high scale
R = ratio between the two scales

If the scale ratio is 2:1, as is often the case, Equation 3-2 simplifies to

$$V_{corr} = \frac{V_L V_H}{2 V_L - V_H} \tag{3-3}$$

The Potentiometer

The need for corrections may be avoided by the use of a potentiometer, provided a galvanometer of adequate sensitivity is used. A typical instrument has a 20-ohm galvanometer with a sensitivity of 20 microamperes per division; assuming that a deflection of $1/10$ of a division can be seen, then the minimum unbalance which can be detected is 0.04 millivolt; this is adequate for most work, except where unusual refinement is sought. The same instrument is frequently used with a 300-ohm galvanometer—excellent for pipe-to-soil potentials—but, as can be seen, inadequate for line current measurement. There is no suitable potentiometer-voltmeter available for this application. Either a potentiometer with proper galvanometer or a high-resistance voltmeter, with corrections for lead resistance, should be used.

Calibration of Test Section

The conductance of a test section may be approximated from the formula

$$K \ (\text{amperes/millivolt}) \ = \ \frac{4 \times \text{Weight (pounds/foot)}}{\text{Length (feet)}} \qquad (3\text{-}4)$$

This is only a rough approximation; the factor may vary from 3.5 to as high as 5.0, and further variation is introduced by the weight tolerance for line pipe. By far the better method is that of direct calibration, as shown in Figure 3-2. Two connections are used for current input, and two others (which will remain as permanent test leads) for the voltmeter connection. Not all of the current flowing through the ammeter, however, flows through the test section; some of it flows through a remote earth path. This effect is quite variable, depending upon coating conductance and earth resistivity; to correct for it, an additional pair of leads is required.

Stray-Current Studies

Whenever fluctuating currents or potentials are observed, stray-current effects are suspected. Useful clues to the source of such strays can be had from 24-hour recordings of line current potentials. If the actual current values are desired, corrections may be made by Equation 3-1; or resistance can be inserted into the circuit so as to make $R_w = R_m$; the correction factor will then be exactly 2 and may be made by selection of the proper chart, so as to record corrected millivolts. Comparison of the charts with records of operation of suspected equipment will often locate the source beyond question.

Long-Line Currents

A use of line current measurement more in vogue in the past than at present is the tracing of long-line currents and the location of the areas of discharge to earth. If currents of this type were the sole cause of pipe line corrosion, all of the anodic areas could thus be found; unfortunately, however, the most common attack on buried pipes is associated with corroding cells of very small dimensions; very often the anode is on the bottom and the cathode on top of the pipe only a few inches away; line current studies cannot locate such a condition.

Cathodic Protection Tests

The current flows associated with cathodic protection systems, either permanent or temporary test equipment, are typically long-line currents; a very valuable part of such testing is in the measurement of these currents; the application will be discussed in a later chapter.

Coating Conductance Measurement

The same thing is true of the determination of coating conductance—a type of measurement which will grow in importance to the corrosion engineer in the future. Line current measurements are much less subject to disturbing factors than are potential measurements and offer the engineer a valuable approach to this difficult subject; this will also be discussed later in greater detail.

Summary

As has already been shown, if test stations along the pipelines have been installed, it is very easy to construct a test section for line current measurement. Although the lengths may vary for these sections, it is advisable to have them at least 100 ft long for 8-inch schedule 40 pipe and as much as 400 ft for 30-inch OD pipe. Again, the multipurpose corrosion meter can be used, since values as low as one millivolt may have to be measured. Also, the test section may be used to measure the coating resistance once the conductance of the pipe itself has been calibrated.

4

Current Requirement Surveys

The Problem: Coated Lines

The simplest approach to the cathodic protection of coated lines is to aim at obtaining a pipe-to-soil potential of 0.85 volt (to a $CuSO_4$ electrode) throughout the line. Meticulous investigations may show that protection can be had at lower figures, but only on bare or very poorly coated structures are the savings likely to be great enough to justify the additional engineering work required.

In determining the amount and distribution of drainage current to produce the desired potential, two basic approaches are possible: (1) the complete protection of the line with temporary installations, these being varied and shifted until a satisfactory combination is found; and (2) the determination of the electrical characteristics of the line, and the calculation of the system from these data.

The first method is almost impossibly complicated in execution and is seldom even attempted. The second is theoretically possible, although in practice a certain amount of trial-and-error procedure is usually necessary. A compromise technique, in which the trial and error of testing the performance of various combinations is done on paper, rather than in the field, is recommended.

Principles of Current Requirement Test

Current is to be drained from the line to a temporary ground bed, and the effects determined for as great a distance as they are measurable (far beyond the point of complete protection); from one or more such tests, the behavior of the line with respect to current drainage can be determined with sufficient accuracy to carry out the design of a complete protective system. The end result of this design will be the specification of drainage units at certain points, draining specified quantities of current; the detail design of anode beds to accomplish this is another matter.

Field Procedure

The following steps are taken, in order:

1. *Static potentials.* A pipe-to-soil potential survey is made over the entire line; permanent test leads are recommended, but bar contacts may be necessary. Spacing may vary from a few hundred feet on poor coating to several miles on good coating; often road crossings will be adequate. Readings should be taken at all casings (reading potentials of both casing and pipe), and on all crossing lines and other structures which might either be shorted or be a source of interference in either direction. Such structures include, in addition to casings, bridges, A-frames, supports of all kinds, and sometimes even such things as metal fences and guy wires.
2. *Polarization run.* At a chosen location, preferably one which appears to be suitable for a rectifier installation, a temporary drainage point is established and a steady current drained for a period of from one to three hours. The pipe-to-soil potential of a nearby point (within a few miles, on a well-coated line; a few hundred feet on a poorly coated one) is observed during this interval to follow the course of polarization. The value of current to be used should be large enough to cover a usefully long portion of the line—perhaps all of it—but should not be too large. The pipe-to-soil potential near the drain point should not, in general, exceed three volts.

3. *Potential survey.* An interrupter, either manual or automatic, is now inserted in the circuit and placed in operation. The schedule should consist of unequal ''on'' and ''off'' periods of predetermined duration; 40 seconds ''on'' and 20 seconds ''off'' is a good combination to use. With this operating, a survey should be made of pipe-to-soil potentials at all of the points covered in 1. *Static potentials.* It is important that the same electrode position be used as was used for the static potential at each location. At each point, both the ''on'' and ''off'' potentials should be read; the ''on'' reading should be taken just before interruption, and the ''off'' reading as soon as possible after interruption. Care should be taken that these two readings are correctly identified; the length of the cycle is used for that purpose. It is *not* safe to assume that the higher (more negative) reading is always the ''on'' reading as this is not the case with a properly insulated lateral or casing. It is recommended that a graph of these values be made in the field as the readings are taken in order to detect any peculiarities which might call for additional readings. Figure 4-1 shows a drawing of an automatic interrupter suitable for use on the above test.

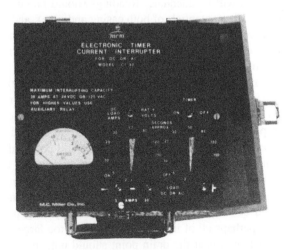

Figure 4-1. Current interrupter. (Courtesy of M. C. Miller Co., Inc.)

4. *Polarization and current check.* The series of potential readings at a point near the drain point, as well as a series of current values, should be continued during the potential survey. It is desirable to keep the current constant throughout the test if this can be done easily. Where this is not possible, it is better for it to change smoothly over a considerable range of values than to change often and abruptly. If the current change is not great, then only one or two values need to be taken during the potential survey. Successful surveys have often been made with only the value at the start and finish (together with those taken during the polarization run). These values should also be plotted in the field.

5. *Line current survey.* If spaced pairs of test leads are available, the line current should be determined on each side of the drainage point, and at a remote location in each direction. Intermediate values may also be useful, particularly if it turns out that there are some short-circuited structures on the line. Like the potential readings, these should consist of "on"-"off" pairs.

6. *Repeat as necessary.* New drainage points are chosen, and the outlined process repeated until the entire line has been covered with values large enough to be usable. It is by no means necessary to achieve complete protection over the whole line, or even over any portion thereof.

Graphical Presentation of Data

All of the information gathered in the survey should then be prepared in the form of a set of curves, as follows:

1. *Current and polarization.* For each test run, the value of drainage current and the potential at the check point should be plotted against time (see Figure 4-2). The curve will have many points close together during the early polarization run, and perhaps only one or two thereafter, but should always be plotted as a complete curve. This will make it possible to interpolate and thus determine the value of the current flowing at any particular time, to correlate with the various potential readings.

2. *Longitudinal distribution curves.* On a single curve, the following should be plotted against distance along the entire line: static po-

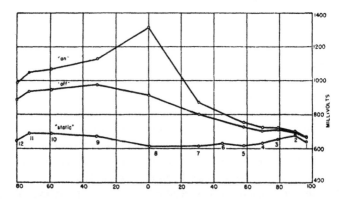

Figure 4-2. Longitudinal distribution curves. Distribution of pipe-to-soil potential along the line; the very obvious difference between the two ends of the line illustrated are attributable to a short-circuited bare casing on the right-hand section. The horizontal unit is 1000 feet.

tential, "off" potential, "on" potential, and line current. There will be only one static curve, but there will be one of each of the others for each test run. Normally, these will overlap, and colors may be required for clear separation. These curves will present most of the data taken in the field in graphic form.

3. *Attenuation curve.* Again on a single sheet (Figure 4-3), but using semilogarithmic paper, the following are plotted against distance (using the same horizontal scale as above):

a. Polarization potential, ΔV_P; this is the difference between the "off" potential and the static potential, at each point.

b. Driving voltage, ΔE; this is the difference between the "on" potential and the "off" potential, at each point.

c. Line current, ΔI; again the value to be plotted is the difference in the "on" value and the "off" value; due attention must be paid to the directions, and the algebraic difference taken. That is, if the test shows 40 milliamperes flowing east during "off" conditions and 300 milliamperes flowing west during the "on" part of the cycle, then ΔI is 340 milliamperes, the net change produced by the test current.

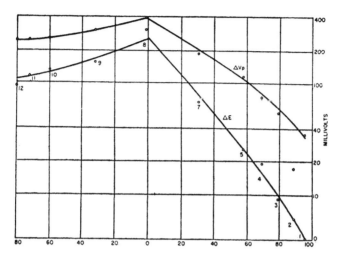

Figure 4-3. Attenuation curves. Driving voltage (ΔE) and polarization potential (ΔV_p) plotted on semilog paper against distance along the line. For an infinitely long line, such curves should be straight lines; for a finite line ending in an insulated joint, concave upward, as on the left-hand section; and for a "grounded" section, concave downward, as on the right-hand section. The test point numbers on all three figures correspond.

4. *Polarization chart* (Figure 4-4). This is usually plotted on log-log paper, although the results are often as good when ordinary linear paper is used. This is a plot of the value of ΔV_P plotted against ΔE, and, where several tests have been run on the same line, or even where there have been repeat tests on the same section with the same or different current values, all pairs from all tests can be combined. In general, it will not be advisable to combine values taken on different lines or on different sections if it is known that pipe size, line age, or type of coating is markedly different. This chart will usually be a "scatter diagram" rather than a smooth curve; often it is advisable to draw limiting lines rather than a single curve through the points. The chart shown in Figure 4-4 is better than most in the degree of conformity to a single curve.

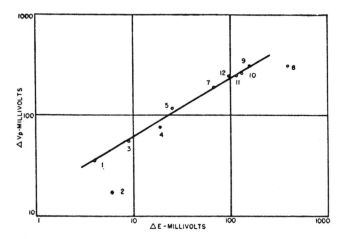

Figure 4-4. Polarization chart. From this plot, the functional relationship between driving voltage and polarization potential may be obtained. Test point 2 is anomalous. The explanation is found in the short-circuited casing at this location, which minimized polarization. The anomaly at point 8 is due to proximity to the anode; the apparent value of the driving voltage is too high.

Current Sources for Tests

Where very large currents—20–100 amperes—are required, a welding machine is the usual source of DC power for current requirement tests. Smaller values can be supplied by storage batteries. It is possible to connect any number of such batteries in series to obtain voltages high enough for medium-to-high resistance ground beds. As high as 120 volts (ten 12-volt batteries) has been used with success.

Very small currents, up to perhaps two amperes, can be supplied with ordinary dry cells, often at a considerable saving in cost, if no storage battery is readily available. This is usually practical only when a low-resistance anode is available. When saltwater is available, as near the seashore, or where a saltwater pit is near the line to be tested, a magnesium anode (even a piece of magnesium ribbon) may serve both as an anode and the source of power. Only where extremely low-resistivity soil or water is at hand, such as saltwater, is this practical, unless the current needs of the line are very small indeed.

Sometimes an adjacent or crossing line, under cathodic protection, or an adjacent insulated section already under protection, can be utilized. In this case it is only necessary to connect a temporary jumper from the line under protection to the one under test, with the interrupter inserted in this connection.

Temporary Ground Beds

To discharge the test current to earth, a temporary anode bed must be found or constructed. Pieces of scrap two-inch pipe, set in six-inch augered holes kept filled with saltwater work well in low- or medium-resistivity soil, provided it is not too sandy or porous. Actually, any metal structure with a large area in contact with the earth may serve, although it must be realized that it is liable to some damage by corrosion. A 10-hour test will remove about 0.36 ounces of steel per ampere of test current.

A fairly common practice, though of doubtful virtue, is that of using another pipe line or another section (beyond an insulating flange) of the same line of an anode. Little or no measurable harm is thus done to a bare line, but on a coated line the damage may be concentrated at a few holidays in the coating. At any rate, whenever this technique is used, the usual ''on'' and ''off'' cycle should be reversed, and one used of, say, 10 seconds ''on'' and 50 seconds ''off.'' In this way the actual exposure is minimized, but it is also true that polarization will be less fully achieved.

Special Conditions

Sometimes fluctuating potentials will be found which make the taking of static potentials all but impossible. These are particularly prevalent on well-coated lines, in high-resistivity soil, and in northern lati tudes. Similar fluctuations are sometimes found associated with thunderstorms—and the stormy conditions may be many miles away on the line under survey, with fair weather at the point of observation. These are temporary, however, and the solution is to come back another day (work around exposed insulated joints is hazardous in thunderstorm weather). But for the conditions associated with earth currents, as described earlier, there is no solution. Care must be taken to check the potential directly across each insulated joint, and the potential between the pipe and each casing, instead of merely checking the two pipe-to-soil potentials involved; but true static potentials cannot be taken, so the design will include a degree of uncertainty not otherwise present.

Summary

We have just seen that the problem of securing adequate cathodic protection of coated lines may be accomplished by using the following procedures for a coated pipe:

1. Install test stations along the pipe line with a few line current test sections included.
2. Take a pipe-to-soil potential difference survey along the pipe before any cathodic protection system is installed. Be particularly careful to get readings of all road crossings where casings are used for protection. If the pipe line goes under a stream, be particularly careful to get all basic information.
3. Select a central location near a position of low soil resistivity where a temporary anode and a direct current source may be installed. (This is a prototype of the eventual cathodic protection system, but at this stage the final details have not yet been determined.)
4. The interrupted tests are now begun, and a potential difference survey of the entire line is made.
5. At the same time a series of potential readings should be made at a point near the drain point where the polarization data may be obtained.
6. Line current and coating conductivity tests should also be run.
7. The data should then be presented graphically in the following curves: (a) current and polarization, (b) longitudinal distribution, (c) attentuation curves (semilogarithmic), and (d) polarization curves (logarithmic).

Current sources should be: (a) a welding machine (if possible), (b) storage batteries in series, and (c) dry cells (rarely).

Ground beds should be: (a) scrap-pipe installed, or (b) unused pipe line (rarely).

5

Rectifier Systems for Coated Lines

General Design Principles

Although the design of a cathodic protection system does not properly belong in a field manual, it is necessary to introduce some of the principles. This is largely true because in the final analysis much of the design must be done in the field. The method described here is based on the survey data as obtained and plotted by the methods described in Chapter 4.

For the most part, rectifier system design involves a trial-and-error method, rather than a straightforward mathematical solution. This is necessarily the case, because in the first place, the mathematical analysis of the attenuation of current and potential along a pipe line is somewhat complicated even with assumptions of uniformity. And in the second place, the choice of practical sites for rectifiers is almost always limited, so the design has to be tailored to fit them.

Attenuation Curves

When current is drained from a single point on a pipe line and discharged to earth, the effects of the drainage—line current, current density on the pipe surface, and pipe-to-soil potential—are all a maximum at the drain point and decrease with the distance from that point. The manner in which they decrease is known as the *attenuation function* of the line. This function is influenced by many factors, among which are the resistance of the pipe (which varies with the weight), the coating, the soil resistivity, connections to other structures, and the method of termination of the line. If pipe weight, coating, and soil resistivity are uniform, or nearly so, and if there are no extraneous connections and the termination is simple, then the attenuation function is a relatively simple mathematical expression; complications in these factors result in complications in the function.

Since the mathematically ideal attenuation curve for a long line is exponential, it is customary to plot all attenuation curves on semilogarithmic paper, on which an exponential curve plots as a straight line. The horizontal scale, which is not logarithmic, represents distance along the line, in units of miles, feet, or thousands of feet, as convenient (see Figure 5-1). The vertical scale is used to plot ΔE (driving voltage) and ΔI (line current.) Note that what is plotted here is the *change* in potential or current when the test current is interrupted. Stray currents, static potentials, and the effects of polarization are all eliminated from the plotted values.

Lines may be classified, insofar as their attenuation behavior is concerned, into four mathematical classes: (1) very long lines, (2) long lines, (3) short lines, and (4) very short lines. These classes are not sharply divided, but merge gradually into one another; and it must be admitted that the classification is rather arbitrary. Before defining these cases, it is well to take a look at the one which is mathematically the simplest of all—the infinitely long line.

If current is drained from a uniform, infinitely long line at a single point and discharged to earth through an infinitely remote anode, both ΔE and ΔI (as well as Δi, the cathodic current density) will be perfect exponential functions, the curve of which, on logarithmic paper, is a straight line. These are illustrated in Figure 5-2. It is important to note that the slopes of the ΔE and ΔI curves are the same; the Δi curve (not shown) would have this same slope. In all of the discussions to follow, the Δi curve, which is seldom used in practice, always has the same shape as the ΔE curve. This is always true unless pipe diameter changes.

In an actual case, if current is drained from a very long line, the attenuation will be essentially as just described. It may be such a good

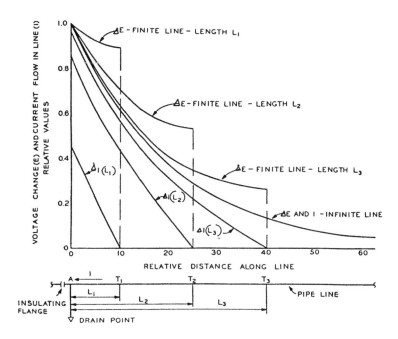

Figure 5-1. Attenuation curves—single drain points.

approximation to a straight line that the instruments used cannot tell the difference. In the case of a line with excellent coating, this is usually the case, even if there is considerable variation in the soil resistivity along the line. A *very long line,* then, is one which behaves, as far as our instruments can tell, like an infinite line. As far as they can be traced, both ΔE and ΔI follow essentially straight lines, and both curves have the same slope.

A *long line* is one which starts out like an infinite line—both curves straight, both with the same slope—but which shows a definite curvature in the lines before the effects of the test become too small to measure. The curves in Figure 5-3 show this characteristic. If the measurements extended only over the part of the line indicated by the two

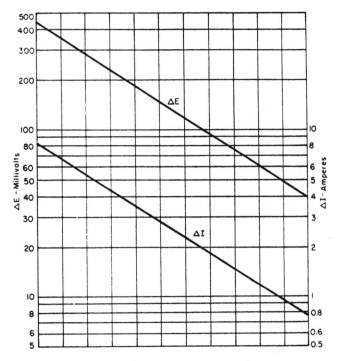

Figure 5-2. Attenuation curves for infinite line. The rate of decline of driving voltage (ΔE) and line current (ΔI) with distance from the drain point. The horizontal scale represents distance along the pipe line; each division might represent as little as 500 feet (large bare line in low-resistivity soil) to as much as 25,000 feet (very good coating).

arrows, it would be difficult to tell it from an infinite line. However, to qualify for the category of *long line,* it must be so long that the potential at the end is not detectable or not large enough to be useful.

A *short line* is one whose length is such that the results at the end are large enough to be useful. It is admitted that the definition makes the length of the line depend upon the sensitivity of the instruments used, but this corresponds to the facts in the case. Thus, a line which is "long" to one instrument may be "short" to another.

Another way in which a *short line* will usually differ from a long one is that the initial slope of the ΔE and ΔI curves will be different. The left-hand portion of Figure 5-4 may be taken as representing a short

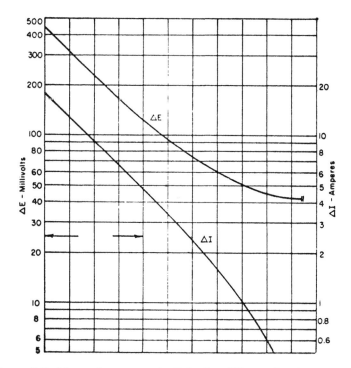

Figure 5-3. Attenuation curves for finite line. When a line is terminated in an insulated joint, the curves take the form shown here. The driving voltage becomes horizontal at the end, while the line current goes to zero. It may be observed that, with a proper choice of vertical scale, either of these curves may be expressed as the derivative of the other. If ΔE and ΔI be interchanged, the above curves apply to a line terminated in a "zero" resistance to ground.

line, and it can be seen that the two slopes are not quite the same; in still shorter lines they are even more different.

A *very short line* is one which is so short that there is little or no difference, or no useful difference, in the potential at the drain point and at the end. In other words, ΔE does not change much from end to end. This means that the line is behaving essentially as would an isolated piece of metal in the soil, all at about the same potential. It is also true of such line that the current, ΔI, is too small to be measured accurately (this is true only for a coated line); it may be measured in the drainage circuit, with an ammeter, but not in the line itself, by IR drop.

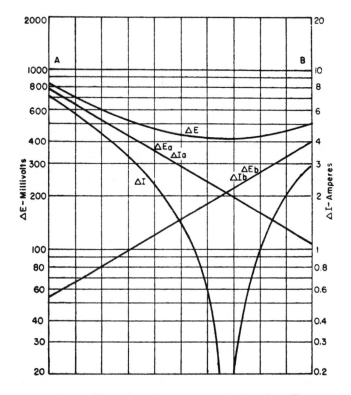

Figure 5-4. Attenuation curves between two drain points. The portion of the above diagram lying on either side of the point where the two straight lines cross, the driving voltage is at a minimum, and the line current is zero, is mathematically identical with the curve shown in Figure 5-3. In other words, an insulated joint could be inserted in the line at the point of zero line current without affecting the behavior at any point.

Line Termination

It does not make any difference how a *very long line* is terminated. If a *long line* is terminated in an insulated joint, there will be a marked divergence in the ΔE and ΔI curves near the end, as indicated in Figure 5-3. The same is true of a *short line*, the difference being only in the magnitude of ΔE value at the end. If either of these categories is terminated by a connection to a massive bare grounded structure, such as a tank farm or a refinery, the ΔE curve will go to zero, while the ΔI curve will become horizontal. A *very short line* must be insulated in order to function as such; if it is grounded, its characteristic cannot be determined.

It is interesting to observe that in the case of a *long line,* where the initial slopes of the two curves are the same, the value of ΔE at the insulated end is exactly twice the value obtained by projecting the original straight line. For *short lines,* if there is not much difference in the slopes, this is nearly true; it becomes less true with shorter lines.

Anode Proximity Effects

In an actual test of a cathodic protection installation, as compared with the mathematical ideal described, the anode will be at a finite distance from the line. The IR drop in the soil around the anode will then affect the measured potentials along the line, and the values of ΔE near the drain point—or, more accurately, near the anode—will be too high.

This shows itself as a localized "peaking" in which there is a distinct departure from the theoretical curve to be expected. In cases where the drain point is not directly opposite the anode, this peak shows up opposite the anode. For design purposes, the value of ΔE_o should not be taken as the actual value measured at the drain point but should be obtained by extending the curve back to that point.

Attenuation with Multiple Drain Points

The distribution of potential and current along a line with two or more drain points is determined by the superposition of the individual curves resulting from each point considered singly. The actual ΔE at any given point is the sum of the ΔE's from all of the drains, and the actual ΔI is the sum of all the ΔI's; in the latter case due attention must be paid to the direction of current; if ΔI flowing north is considered positive, then ΔI flowing south is to be taken as negative. Figure 5-4 shows a section of pipe line lying between two drain points, the ΔE and ΔI curves from each of them, and the resultant ΔE and ΔI for the actual case.

In plotting these composite curves, it should be noted that the numerical values of current and potential are added, not the logarithms; that the absolute value of the current is used, without regard to sign; and that the conditions that the slopes of the two curves are equal where they coincide or can be made to coincide by proper choice of scales no longer holds, as it did for curves from a single drain point.

Design Procedure

The application of these curves to the actual problem of design proceeds as follows. A suitable site for a rectifier is chosen, using the criteria of power availability, low soil resistivity and accessibility, and an assumed value of the drainage current at that point is chosen. The distribution of current and potential along the line from this drainage is then projected, using the test data obtained. The actual pipe-to-soil potential at any point may be predicted by adding the static potential (from the survey) plus the driving voltage (from the attenuation curve) plus the polarization potential (from the ΔV_P vs. ΔE curve).

It may be that the entire section under consideration can be protected adequately from the single drain point assumed. If not, then another must be chosen, and attenuation curves prepared for it also. Then the section between the two is studied, using the method of superposition. For any given point, the pipe-to-soil potential is predicted by adding the static potential, plus the *sum* of the two driving voltages, plus the polarization potential attributable to this composite driving voltage. When, by trial and error, a combination of drain points has been found which places the entire line under protection, a system has been found which satisfies the conditions. Whether it is the most economical system which will accomplish this result cannot be told except by comparing it with other possible combinations for initial cost and for operating cost.

An Alternate Method

A somewhat less accurate but faster method may be used for initially sizing the rectifiers for a planned pipe line cathodic protection system. This involves using the basic data already listed as necessary for a pipe line calculation:

1. *Coating conductance,* g. This may be determined using the line current setup as described in Chapter 3, and the method described

later in Chapter 12. Certain values incorporating the resistivity of
the pipe line itself and varying wrapped coatings are as follows:

a. 1–10 micromhos/sq ft. Excellent coating in high-resistivity soil
b. 10–50 micromhos/sq ft. Good coating in high-resistivity soil
c. 50–100 micromhos/sq ft. Excellent coating in low-resistivity
soil
d. 100–250 micromhos/sq ft. Good coating in low-resistivity soil
e. 250–500 micromhos/sq ft. Average coating in low-resistivity
soil
f. 500–1000 micromhos/sq ft. Poor coating in low-resistivity soil

Consequently, even though the conductance tests have not been
run, it is possible to estimate this value of g.

2. ΔE *at low point on the pipe line,* ΔE_x. Criterion 3 mentioned in
Chapter 3 may be used for the design assumption. That is a potential difference change of -0.3 volt. This is enough to raise the
average static value of coated steel in soil of about -0.55 v to a
Cu-CuSO$_4$ electrode to -0.85 v, enough to maintain the ap-
proved cathodic protection criterion.

3. ΔE *at drain point,* ΔE_A. A good practice is to limit the voltage
change at the maximum point to around 1.5 volts. This is to pre-
vent disbonding of the coating. In certain circumstances this value
may be exceeded, but it is better to use the conservative value in
the preliminary estimation.

4. ΔI *at drain point,* ΔI_A. This is the value to be determined. It may
be calculated from the following relationship determined from
one of the two modified attenuation curves (Figures 5-5 and 5-6):

$$I_A - \text{amperes per inch} \times \frac{\Delta E_x}{0.3} \times D$$

where D is pipe OD in inches.
For the special case where $\Delta E_x = 0.3$ and $\Delta E_A = 1.5$,

$$\frac{\Delta E_A}{\Delta E_x} = 5.0 \text{ and } \frac{\Delta E_x}{0.3} = 1.0$$

5. *Method of calculation using Figures 5-5 and 5-6.* If g, the coating conductivity in micromhos per square foot, is known or estimated and the relationship $\Delta E_A/\Delta E_x$ or $\Delta E_A/\Delta E_T$ is assumed, it is possible to calculate both I_A and the distance it will protect for either an infinite pipe line (using Figure 5-5) or a finite pipe line (using Figure 5-6).

Example A: Infinite Line

A pipe line has an OD of 30.0 inches and a coating conductivity of 100 micromhos per sq ft. What is the current change at the drain point, and how far will this current protect an infinitely long pipe line? Assume $\Delta E_A = 1.5$ and $\Delta E_x = 0.3$. Then, from Figure 5-5, $L = 30.0$ and $\Delta E_A/\Delta E_x = 5.0$

$$\left(\frac{I_A}{D}\right)\left(\frac{0.3}{\Delta E_x}\right) = 0.70$$

where:
distance on actual line $= 30,000$ ft,
$I_A = 0.7 \times 30.0 = 21.0$ amps.
Therefore, a rectifier developing 21.0 amperes will protect 30,000 feet (or about 5.7 miles) in either direction from the drain point (Figure 5-5).

Example B: Finite Line

A flow line 20,000 ft long and nominal 3-inch pipe insulated from the well by an insulating flange at the Christmas tree and by another insulating flange at the tank battery has a coating conductivity of 500 micromhos per square foot. (1) What is $\Delta E_A/E_T$? (2) What is ΔI_A required?
From Figure 5.6,

$$\frac{\Delta E_A}{\Delta E_T} = 5.8 \text{ (If } E_T = 0.3, \text{ then } E_A = 1.74.)$$

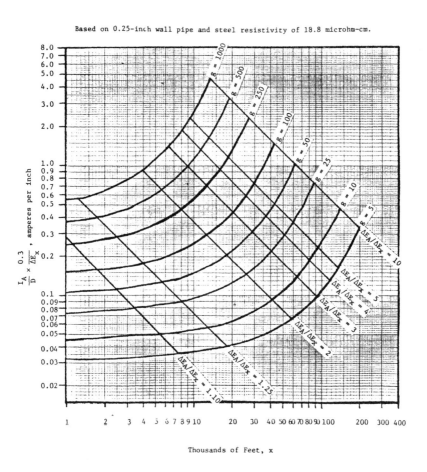

Figure 5-5. Drainage current vs. distance coated infinite line.

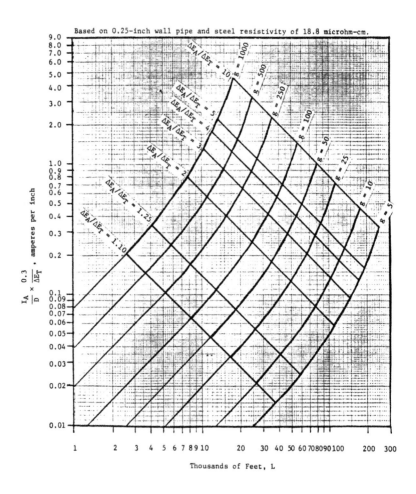

Figure 5-6. Drainage current vs. length coated finite line.

$$\frac{I_A}{D} \times \frac{0.3}{E_T} = 1.8 \text{ amperes/inch}$$

$I_A = 1.8 \times 3.625 \text{ inches} = 6.525 \text{ amperes}$

Therefore, as shown in Figure 5-1, the attenuation curves for the two examples are superimposed on a hyperbolic curve showing voltage and amperage differences.

The Economic Balance

There are far too many factors influencing the choice of the optimum design to do more than barely outline the principles, particularly in a field manual. The lowest *initial cost* will be one which uses the smallest number of drainage points, each of relatively large size, limited usually by the maximum permissible pipe-to-soil potential. The lowest *operating cost* will be one which uses a larger number of smaller drain points at much closer spacing, thereby avoiding the "wasted" power used in over-protecting sections near the units. Somewhere between these extremes will be a system which is most attractive from every point of view—first cost, operation, maintenance, and supervision.

The final choice is influenced by factors quite remote from field conditions. Factors such as availability of capital, the earning rate of the company, and the tax and income position. Ultimately the decision is one for management, with the corrosion engineer furnishing the technical analysis upon which it is to be based.

Summary

Two methods may be used. First, use all the data accumulated in the tests in Chapter 4 to do the following:

1. *Select rectifier sites.* The availability of power lines has to be considered, but also the location of low-resistivity areas for the accompanying anode beds.
2. *Calculate the attenuation curves.* All the data should give some indication of the ΔE and ΔI to be found. Also calculate for (a) infinite, and (b) finite if the line is short or is to be sectionalized by insulation flanges.

3. Plot these curves for multiple drain points if more than two rectifier sites are to be chosen.
4. Compile all these data together to size the rectifier. Obviously, by now we have the following data:

g = coating conductivity, micromhos per sq ft.
R = soil resistivity, ohm-cm.
ΔE = at drain points and midpoints, volts.
ΔI = at drain points, volts.

The second method is a check, but one which may be of use:

1. Use the same g, R, ΔE, and ΔI available from the preliminary data. If these are not available, estimate them from commonly accepted standards.
2. Now use average attenuation curves similar to those first used by Chevron Oil Company in the 1950s. These are Figure 5-5 for infinite and Figure 5-6 for finite lines.

See the test problems for examples.

6

Ground Bed Design and Installation

Design Principles

The ideal design for a cathodic protection system is the one which will provide the desired degree of protection at the minimum total annual cost over the projected life of the protected structure. Total annual cost means the sum of the costs of power, maintenance, and charges against the amount of capital invested. Both the operating cost (power consumption) and installation cost are influenced by the resistance of the anode bed. This quantity occupies a central role in the design of any impressed current cathodic protection system.

Note that it is not the lowest-resistance ground bed which is the best, nor is it the one whose installed cost is the least, nor is it yet the one whose power consumption is the least. It is, rather, the one whose resistance is such as to fit into an overall system whose total annual cost is the least.

At one stage in the design process described in the preceding chapter, it has been decided—tentatively at least—to drain a certain amount of current from the line at a certain point. It is then necessary to design a ground bed and anode combination which will do this for the least annual cost. The first choice to be made is the general type of anode to be used. In most cases this will be a row of equally spaced vertical anodes; other types, and the special conditions in which they are likely to be better, will be discussed later. Having chosen the type—if the row of verticals is selected—the next item to be determined is the hole diameter and the spacing; and then, finally, the total number of anodes to be used.

Among the types of vertical anodes in common use, perhaps the most popular consists of a graphite rod, usually 3 inches in diameter and 60 inches long, centered in a cylindrical column of well-tamped coke

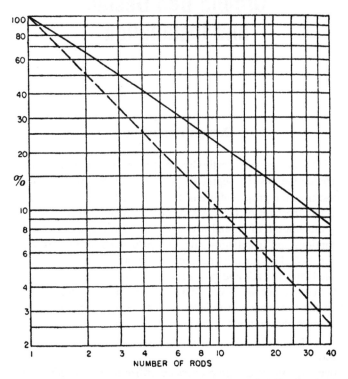

Figure 6-1. Resistance of a number of rods in parallel. Curve shows the total resistance to earth of a number of 3-inch by 60-inch graphite anodes, each installed in 10 feet of tamped coke breeze backfill 8 inches in diameter, when connected in parallel in a straight line at 20-foot spacing. The total resistance is expressed as the percentage of the resistance of a single rod. It is assumed that the soil is of uniform resistivity; if the resistivity increases with depth, the curve should be raised; if it decreases, it should be lowered. The dashed line shows the parallel resistance which would be obtained if there were no interference.

breeze backfill (see Figure 6-1). The column of coke is the actual anode; the rod serves merely to establish and maintain contact. Column diameters range from 8-inches to 16-inches, occasionally even larger; 10-inches is perhaps the most common. Depths used range from 8 feet to 12 feet in general, with 10 feet a very common value. Entirely typical, then, is a 3-inch by 60-inch graphite anode in a 10-inch by 10-foot hole, with the backfill column extending to within 2 feet of the surface.

The resistance to earth of a single anode of the type described is given by the expression

$$R = \frac{\rho}{2\pi L} \ln \frac{4L}{ae} \qquad (6\text{-}1)$$

where:

R = resistance, ohms
ρ = soil resistivity (assumed constant to great depth)
L = length of the coke breeze column, cm
a = radius of the column, cm
e = base of natural logarithms, $e = 2.718\ldots$
ln = natural logarithm
This can be simplified to

$$R = .012 \frac{\rho}{L} \log \left(\frac{35L}{d}\right) \qquad (6\text{-}2)$$

where:

R = resistance, ohms
ρ = soil resistivity in ohm-cm (again assumed constant to great depth)
L = length of the coke breeze column, feet
d = column diameter, inches
As indicated by the "log," the common logarithm is used.

When these formulas are applied to the dimensions given, coke breeze column 10 inches in diameter by 8 feet deep—the results (by either formula) are R = .002 ρ, or $R = \rho/500$. This is a good rule-of-thumb approximation to remember for the resistance to earth of a single vertical anode.

When several anodes are connected in parallel, the resistance of the whole group is somewhat greater than the value obtained by dividing the resistance of one anode by the number of anodes. This is due to the "interference" between adjacent anodes, or to the crowding of the current paths in the earth so that the current density, and hence the total voltage drop, is greater than it would be around a single anode. When the anodes are set in a straight line, at equal spacing—the usual arrangement—the total resistance of a group may be calculated by the following expression:

$$R_n = \frac{\rho}{\pi n} \left[\frac{1}{2L} \ln \left(\frac{4L}{ae} \right) + \frac{1}{S} \left(\frac{1}{2} + \frac{1}{3} + \frac{1}{4} + \ldots + \frac{1}{n} \right) \right] \qquad (6\text{-}3)$$

where n is the number of anodes, and S is the spacing between them. All dimensions are in centimeters, as in the earlier expression.

All of these formulas are based on the assumption that the soil resistivity is uniform to great depth; this is in fact very seldom the case, and it is often difficult to decide just what value of resistivity to use. For a single anode, it is usually quite accurate to use a value corresponding to a pin spacing (four-terminal method) equal to the depth to the center of the anode column. In the case described, the value of six feet would be used.

For multiple anodes, however, the situation is different. The interference or crowding effect depends upon the resistivity of the soil at greater depths. Reasonably good results will be had if a value is used corresponding to a pin spacing equal to the spacing between anodes. To use these two different values in the same formula, it is necessary to perform some algebraic operations on it so as to separate the two parts. If, at the same time, constants are introduced permitting us to use conventional units instead of cm and to use common instead of natural logarithms, the following expression is obtained:

$$R_n = .012 \frac{\rho_1}{nL} \log \frac{35L}{d} + \frac{\rho_2}{S} F_n \qquad (6\text{-}4)$$

where:

R_n = resistance of n anodes, ohms
ρ_1 = resistivity at a spacing equal to the depth of the center of the anode, ohm-cm
L = length of the anode (coke breeze column), feet
log = common logarithm

d = diameter of the coke breeze column, inches

ρ_2 = resistivity at a spacing equal to the anode spacing, ohm-cm

S = anode spacing, feet

F_n = expression involving the series of fractions, with certain constants included. (This function is tabulated for convenience in Table 6-1.)

By using these expressions, it is possible to calculate the resistance of any number of anodes installed in a specific location. For the usual 3-inch by 60-inch graphite rod, it is customary to impose a maximum current limitation of 3 amperes per rod, in the normal coke breeze backfill (see later discussion of bare rods). For a given drainage point, this determines the minimum number of rods to be used, i.e., if the current drain is to be 15 amperes, then at least 5 rods will be required, regardless of resistance values.

The resistance for this minimum number of anodes is computed, and for several larger numbers. With the costs of the various materials used, and the costs of ditching, boring of holes, and other components of the installation, it is then possible to plot a curve showing the total installed cost of a ground bed at this particular site as a function of the number of anodes used (curve A in Figure 6-2). As the number of rods used is increased, the resistance of the total bed decreases. This means that the

Table 6-1
Interference Factor F_n

n	E_n	n	F_n
2	.00261	16	.00155
3	.00290	17	.00150
4	.00283	18	.00145
5	.00268	19	.00140
6	.00252	20	.00136
7	.00238	21	.00132
8	.00224	22	.00128
9	.00212	23	.00124
10	.00201	24	.00121
11	.00192	25	.00118
12	.00183	26	.00115
13	.00175	27	.00112
14	.00168	28	.00109
15	.00161	29	.00107

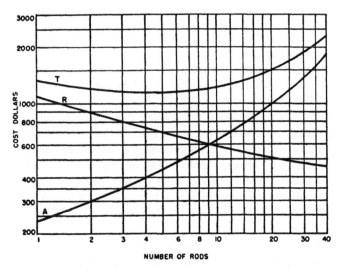

NUMBER OF RODS

Figure 6-2. Total installed cost. Curve A shows the installed cost of an anode bed (as described in the text) as a function of the number of rods; Curve R shows the installed cost of the rectifier required for each number of rods; Curve T shows the total installed cost—the sum of curves A and R—as a function of the number of rods in the bed (1983 prices).

voltage needed for the rectifier to discharge the required current decreases. This voltage may be taken as $V = I(R_n + 2)$; the added 2 volts is for the galvanic difference between a graphite anode in coke breeze and the steel pipe in soil. From this information, and a price list of the type rectifier to be used, a curve can be plotted showing the installed cost of the rectifier as a function of the number of anode rods used. This is shown as curve R in Figure 6-2.

This latter curve may sometimes be derived more easily by the use of an intermediate curve, an example of which is shown in Figure 6-3. The DC wattage of a rectifier is the product of its rated DC voltage and current, i.e., $W = EI$. For a given style and make of rectifier, the purchase price and the total installed cost in a given location are functions of the wattage. Normally, this will not be as smooth a curve as shown in the

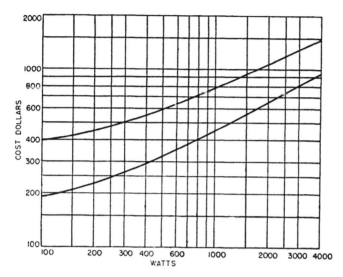

Figure 6-3. List price and installed cost of rectifiers. The lower curve shows the approximate list price of oil-immersed selenium rectifiers as a function of the DC rating in watts (volts) times amperes). The upper curve shows the estimated installed cost under an assumed set of conditions. Curves should be prepared for each estimating job (1983 prices).

example, for a variety of reasons, but such a curve can be plotted from the supplier's price list and the estimated costs of installation.

Returning to a consideration of Figure 6-2, the curve T is the sum of the other two and shows the total installed cost of the complete installation as a function of the number of anode rods to be used. The annual charges against the investment will be directly proportional to this curve. The exact percentage to be used is a figure which must be obtained from management, as it is dependent upon the company's tax position, cost of financing, earnings, regulatory conditions, and a number of other factors which are far removed from field engineering. In the example used here, this figure has been taken as 15%. This is probably too low, as actual values in current use by most companies range from 20% to as high as 50%. The curve I in Figure 6-4 is the annual cost curve, being 15% of curve T in Figure 6-2.

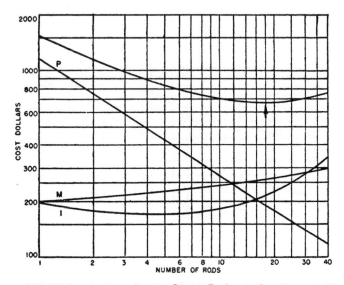

Figure 6-4. Minimum annual cost. Curve P shows the power cost per year curve I the investment charges per year, and curve M the yearly maintenance cost. The upper curve is the sum of these three, and thus shows the total annual cost as a function of the number of anode rods installed. A ground bed of 18 rods is seen to be the most economical under the assumed conditions (1983 prices).

Curve M shows the expected annual maintenance cost for the installation. This curve is based on the company's experience but should take into account remoteness and similar factors. It will be noted that there is little difference between a large anode bed and a small one; the rectifier takes most of the inspection and maintenance costs.

Curve P shows the annual power cost for the operation of the unit. In any actual case, since power rates vary stepwise, this curve would show breaks in slope. In fact, since there is usually a minimum monthly bill, the left-hand portion of the curve would slope as shown, but past a certain point the curve would be horizontal; there would be a saving in actual power consumption, but not in the monthly power bill. This curve should always be drawn for the specific case, taking actual rates into account, and also taking into account whether or not the rectifier is the only load on the particular service installation. Rectifier efficiency can usually be assumed to be about 50%, so the power consumption should be about double the DC wattage as computed previously.

The upper curve in Figure 6-4, without a designation, is the sum of the other three curves—and is the one for which all the others have been made. It shows the total annual cost of the installation as a function of the number of anode rods installed, and it tells the number of rods which will give the most economical installation. In the example shown, this number is 18, but it can clearly be seen that there is not much difference from about 12 to 25. This is correct, it should be noted, only if each number of rods is properly matched to just the right size rectifier.

Disturbing Factors

The actual process is in some respects more difficult, and in others easier, than the description given earlier. The thing which often makes it much easier is experience—often only a small part of the process described actually has to be done. Complexities and difficulties are introduced by nonuniformities and by the fact that few of the functions are continuous ''smooth-curve'' functions as illustrated, but tend, rather, to be broken lines.

Power cost has already been mentioned in this regard; rectifier cost is another. For a certain well-known brand of rectifier, for example, it will be found that there is only a slight increase in price for increasing voltage ratings (for the same current) up to 28 volts. There is then a big jump from the 28-volt unit to the 32-volt unit, and again only a modest increase up to 56 volts, where another jump is encountered. The reason for this is that the voltage rating of a single selenium stack is 28 volts, so that one has to be used for any rating up to that value. Above that, two must be used in series, and so on.

In the example given, only one size and spacing of anodes has been considered; obviously there are many possible combinations. To investigate these thoroughly would require repeating the process several times and choosing the best of the various combinations tried.

Above all, there is the immense variability of soil resistivity. The formulas given show how to design—with a reasonable degree of accuracy—when the soil varies with depth. But no methods are given for dealing with horizontal variation, which is certainly common enough.

When recognition is then given to the fact that resistivity also varies with time, because of moisture variation, it is seen that the calculation of ground bed resistance is not a highly accurate technique. Large deviations from the calculated value are to be expected, and must be allowed for. In addition, it is wise to provide excess capacity, especially when a new line is being protected, to allow for future coating deterioration.

Field Modification

One of the best methods of perfecting such a design, however, is that of field modification while the installation is in progress. The connection to the pipe line itself should be made, and the anode lead laid out along the line of the proposed anode field. Then as the anodes are installed, they should be connected to the lead—a temporary connection may be used—and the total loop resistance between the pipe and the anodes should be measured, the anodes being connected one at a time. A convenient way to do this is to use, successively, one, two, and all three cells of a storage battery, plotting the current values measured with an ammeter and determining the resistance by projecting the line thus obtained back to zero current; the slope of the line gives the resistance, and the voltage shown for zero current is the galvanic potential between the graphite and the steel.

As these values are determined for successive numbers of connected anodes, a plotted curve will show the trend, and Figures 6-2, 6-3, and 6-4 can be replotted in the field. Changes can thus be made in the number of anodes installed while the equipment and personnel are available.

Another technique of field modification (see Figure 6-5), somewhat less perfect than the one described above, but usually leading to a design which is not far from the true minimum, is as follows: make the pipe connection and install the ground bed, complete, as designed; then measure the loop resistance by the battery method described and purchase a rectifier of the voltage necessary to deliver the desired current through the ground bed as built. This requires, of course, a delay of some weeks, since the rectifier is not to be ordered until the ground bed has been completed; nevertheless, it has the undisputed advantage that a workable system will be had, even if the resistance departs markedly from the predicted value.

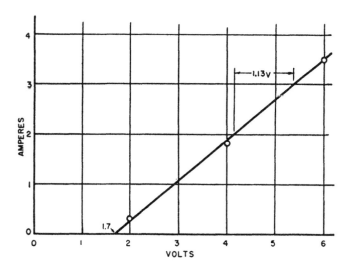

Figure 6-5. Anode resistance by battery test. One, two, and three cells of a fully charged storage battery are connected in turn between the pipe connection and an anode or group of anodes. The slope of the line gives the resistance; in this case, it is 1.13 volts/ampere, or 1.13 ohms. The voltage indicated for zero current, 1.7 volts, is the galvanic potential between the anode and the pipe line.

Installation Methods

Figure 6-6 shows an installed vertical anode, in cross section. The principal points to be observed in the installation are:

1. The coke breeze must be *thoroughly* tamped; loose backfill can give disappointingly high resistances and shorten anode lives.
2. Buried connections must be protected with extreme precautions against the entrance of any moisture, for any discharge of current to earth from the cable will destroy it in a matter of days or hours.
3. Care should be taken to protect the cable connection to the anode; this is the weak point in all known graphite anodes, and even if the joint does not fail from rough-handling, the tiniest crack will permit the entrance of moisture, with inevitable failure.
4. Burial should be at a sufficient depth (18–24 inches) to protect against accidental damage. It must be remembered that the anode

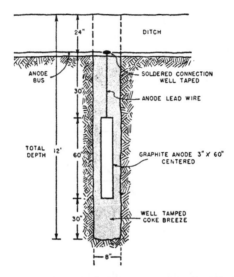

Figure 6-6. Typical single graphite anode vertical installation. The critical items are the coke-breeze tamping, the protection of the anode lead connection, and the careful protection of the connection to the anode bus.

lead may be severed by electrolytic action if there is the slightest break in its coating. The alternative horizontal anode installation is shown in Figure 6-7.

Consideration should always be given to the possibility of placing the anode lead above ground. Often it may be strung on existing poles, on fence posts, or on short posts erected for the purpose. The voltage is so low that little insulation is needed, and the shock hazard is nil. Bare cable may be used if it can be supported above ground, at a considerable saving. However, large copper cable is very attractive junk, and buried installations are preferred if there is a theft danger.

High-Silicon Iron Anodes

As mentioned, the entire coke breeze column acts as the anode; the central rod is merely a means of establishing an electrical connection to

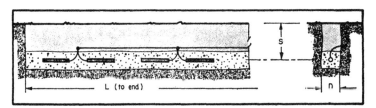

Figure 6-7. Horizontal anode. Resistance is calculated on the basis of the total coke breeze backfill, which is the actual anode. The graphite or high-silicon iron rods serve merely to establish good electrical contact.

the backfill. If this were always true, then a simple piece of iron rod or pipe could be used instead of the more expensive graphite. Actually, the current leaves the graphite in two different ways. Much of it flows by direct conduction to the backfill and does no electrochemical damage to the graphite in so doing; but a small part flows to the moisture which penetrates to the core. This portion of the current, which flows electrolytically instead of by direct conduction, does attack the graphite, but not to an intolerable extent.

It is important to note, however, that steel under these conditions will be severely attacked, and also that graphite without backfill will be severely attacked. The significance of the latter fact is that when graphite is used under conditions which make the installation of properly tamped backfill difficult or impossible, trouble is encountered.

For example, if the hole fills up with water and cannot be pumped dry, correct tamping is impossible; graphite will be attacked, and the installation will have a disappointingly short life. If the hole caves badly, there will be large inclusions of soil next to the anode, and the graphite will be attacked at those points. And, finally, if there is quicksand—really just the extreme of the two conditions mentioned—it will be completely impossible to install tamped backfill without recourse to expensive casings or similar measures; and bare graphite will be severely attacked unless current is limited to about one-fourth the usual value.

One answer to these problems is in the use of high-silicon cast-iron anodes. These are almost inert under most conditions and will normally show only small weight losses—far below the 20 pounds per ampere year for steel. For ground bed applications, the most common size is 2 inches by 60 inches. When installed in the same way as the graphite anode shown in Figure 6-6, it will function just as does the graphite, but it will be almost impossible to tell the difference by any test. These anodes are commonly known in industry by their commercial names, "Duriron" for the 14% Si-cast iron alloy and "Durichlor" for installation in chloride-containing soils.

If the installation is in a difficult situation so that proper tamping is not obtained for one reason or another, the high-silicon iron anode will not show the severe attack which characterizes the graphite at those areas which are in contact with caved-in soil. Or, if there has been an invasion of water to the extent that the whole job of tamping is poor, again the high-silicon anodes will not show excessive attack. Finally, if augering and tamping is completely impossible, as in the case of quicksand, these anodes may be used entirely bare. Naturally, the resistance will be much higher per anode (because of the smaller dimensions) so that ordinarily more anodes will be needed than would be the case if backfilling were possible. One problem with the high-silicon anodes is that, like cast iron, they are extremely brittle and must be packed and handled with care. Backfill is less necessary than with graphite.

Steel Anodes

When the soil resistivity is high, the number of anodes needed to secure an economically low resistance may be impossibly large. In this case, attention should be given to the possibility of using scrap steel pipe or some similar available scrap as an anode. In this way the very large area needed may be obtained, and the high resistivity will make the current density so low that an economically feasible life can be had. Such a ground bed, in high-resistivity soil, can be much less expensive than those using vertical rods described earlier. The dividing line lies somewhere between 5000 ohm-cm and 20,000 ohm-cm; only a set of calculations for the specific case can determine its point.

Such a horizontal anode is usually installed by, in effect, laying a short section of bare pipe line. The joints need not be made by the usual careful welding; in fact, the necessary electrical connection is often made by welding two or more straps of metal from one section to another. These joints should then be coated, to slow down the attack at these points. Several connections should be made to the anode. These can take the form of sections of coated pipe welded to the anode and

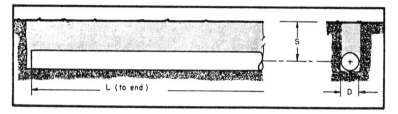

Figure 6-8. Horizontal scrap steel pipe anode.

projected above ground. A "pad" around each should be coated, and some feel that it is good practice to coat a strip down the top of the entire anode. This will tend to raise the total resistance, but it also helps to prevent early segregation and loss of part of the anode. (Figure 6-8 shows a horizontal steel scrap anode installation.)

The resistance of an anode of this kind is given by

$$R = \frac{\rho}{192L}\left\{\ln \frac{4L}{D} + \ln \frac{L}{S} - 2 + \frac{2S}{L}\right\} \qquad (6-5)$$

where:
  R = total resistance, ohms
  ρ = soil resistivity, ohm-cm
  L = anode length, feet
  D = anode diameter, feet
  S = depth to the center, feet
  ln = natural logarithm

The resistance of such an anode may often be reduced by mixing salt and gypsum with the backfill around the pipe, but the total effect of this salting is difficult to determine in advance.

Horizontal Graphite Anodes

Sometimes it is necessary to install an anode in a location where rock is encountered at a shallow depth, or where the soil resistivity increases markedly with depth. A site of the latter type should be avoided if at all possible, but sometimes it can be coped with by a horizontal installation of graphite or high-silicon iron anodes. A ditch is excavated to whatever depth is practical, and a horizontal column of coke breeze is laid therein, usually square in section. The anodes are laid horizontally in the center of this column, with spacing between them up to perhaps twice the anode length. If the coke breeze is well compacted, the entire column will act as an anode. The expression given for the horizontal steel anode is used for the resistance, with D representing the width instead of the diameter.

Deep Anodes

Sometimes the distribution of soil resistivity is the opposite of that just described; there may be high-resistivity soil at the surface and lower resistivity at greater depth. In such circumstances, it may be advisable to install a deep vertical anode in a hole drilled very much like a well. In fact, abandoned wells have been used for this purpose. This hole may be "wet," i.e., full of water, or it may be dry. In holes up to perhaps 100-foot depth, it is possible to tamp coke breeze in a dry hole; at greater depths, and in all wet holes, it is necessary to pour the backfill in. It has been found that coke breeze will, in time, settle quite compactly in a hole full of standing water, so that the results are comparable with those obtained by tamping.

The anodes used in these installations are usually high-silicon iron, both because of relative ease of installation and for their greater immunity to attack when the backfill is less than perfect. A number of anodes are installed in the same hole—depending on the thickness of the low resistivity layer—with the spacing between rods usually about equal to the length of one rod.

A Typical Ground Bed Installation

Problem A: A ground bed for a rectifier system is installed in 1000 ohm-cm soil. Seven 3-inch by 60-inch vertical graphite anodes are to be installed. What is the resistance of the anode bed if the anodes are spaced 10-feet apart? Answer: 0.59 ohms (from Figure 6-9).

Problem B: A ground bed for a rectifier is composed of ten 2-inch by 60-inch bare "Duriron" silicon anodes. The soil resistivity is 3000 ohm-cm. If the anodes are spaced 20-feet apart, what is the anode bed resistance? Answer: Using Figure 6-10, $0.57 \times 3000/1000 = 1.71$ ohms.

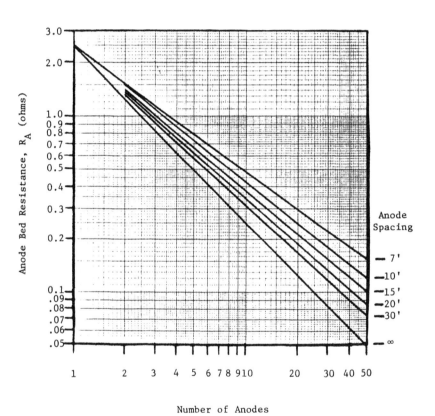

Figure 6-9. Anode bed resistance vs. number of anodes, 3-in. by 60-in. vertical graphite anodes in backfill.

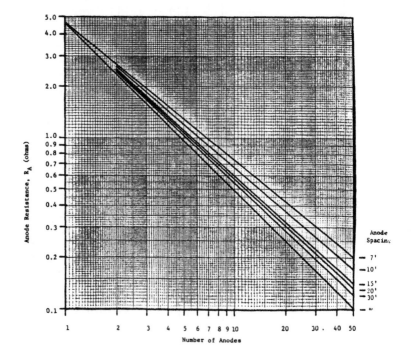

Figure 6-10. Anode bed resistance vs. number of anodes, 2-in. × 60-in. bare vertical anodes (Duriron).

Summary

Once the rectifiers have been sized, the ground beds should be designed. Several anodes are available:

1. Scrap steel.
2. High-silicon iron anodes, called "Duriron" and "Durichlor."
3. Graphite.
4. Special anodes (used mainly for seawater application).

Various arrangements are also available: horizontal, vertical, vertical in parallel, and deep anodes. The calculations for these have been described. The basic formulas, plus others, have been included to show a typical installation design.

7

Galvanic Anodes on Coated Lines

Sometimes, for economic reasons, it may be advisable to use a galvanic rather than a rectifier system for pipe line cathodic protection. Only two metals are presently used for land galvanic anode cathodic protection systems. These are magnesium and zinc.

When and Where to Use Magnesium

The principal advantage of magnesium anodes, as compared to the use of rectifiers, is that the current may be distributed much more readily, since it comes, in effect, in small packages. There is always a certain amount of over-protection of a pipe line near the drain point; this is wasted current. If, therefore, it is possible to control the distribution of current so that each part of the line receives just what it needs, much of this waste will be eliminated. This means that properly applied magnesium anodes can protect a line with less *current* than a system of rectifiers; however, current from rectifiers costs less per ampere than that from magnesium, so it is not always true that a line can be protected for less *cost* with anodes.

It is also true that magnesium anodes are not economical sources of current in soils of high resistivity; their use is limited almost entirely to soils of less than 3000 ohm-cm, except in some special cases where only small amounts of current are needed. It is important to note that, when the protection of a coated line is contemplated, it is only necessary that short sections of low-resistivity soil be found at intervals; and the better the coating, the greater these intervals can be, and still make the use of magnesium feasible.

The field of application, then, of magnesium anodes, is in the protection of bare lines, of isolated structures, as auxiliary protection (to be discussed later), and in the protection of some coated lines, provided the soil conditions are suitable for their use. This latter application will be discussed here, and the others in the next chapter.

When and Where to Use Zinc

The natural potential between zinc and steel is much lower than that between magnesium and steel. Consequently, other things being equal, the current output of a zinc anode will be lower than that of a magnesium anode of the same size. In lower-resistivity soil this may not be a disadvantage; on the contrary, it may make it possible to install anodes for a longer projected life than would otherwise be possible.

At low current output, the efficiency of zinc does not fall off as does that of magnesium. It is therefore possible in many cases to install zinc anodes with a projected life of 20, 30, or even 40 years; magnesium is seldom practical beyond 10 years. Tying up capital in such a long-range program is not always the practical thing to do, but in cases where access and installation are difficult—as with gas distribution systems in congested urban areas—this is often the ideal solution.

In some circumstances the low driving potential of zinc with respect to steel can be turned to advantage. A generous set of zinc anodes will raise a steel structure to about 1.1 volt with respect to a copper sulfate electrode, but cannot raise it higher, since this is about the potential of zinc itself. If, then, the coating fails in part, or if the resistivity of the medium changes, as with tidal water, the current output of the zinc anodes will vary over a wide range, with very little shift in the potential of the protected structure. It can thus function essentially as a constant potential system, and do so at a level which is quite appropriate for steel (magnesium can do the same thing, but at the much higher potential of 1.6 volts, which may require four times as much current).

General System Design

A fact which might be faced is that it is virtually impossible to design on paper an effective and economical magnesium anode system for the complete protection of a coated line. If the current requirements of the line were uniform throughout its length, and if the soil conditions were also uniform, then such a design might be possible; but neither of these conditions is ever found in practice. The outstanding advantage of magnesium in this application is the flexibility which makes it possible to distribute the current where it is needed; and only by taking full advantage of this characteristic is it possible to achieve economy in its use. It might seem that a survey could be made which would supply the data needed for the design of a system having proper distribution; such a survey would have to determine the current requirements of various line sections, and the soil resistivities associated with them. A survey in sufficient detail to make the design possible would be an expensive project; a much more reasonable technique is to use the partially completed installation as a continued survey, and obtain the data as it is needed; in other words, tailor the system to fit, in the field.

Some kind of preliminary estimate is necessary. It may be assumed that about 2 milliamperes per square foot of equivalent bare metal will be required to afford complete protection, as indicated by the criterion of 0.85 volt to a copper sulfate electrode. The question to be decided is the effective bare area of the coating. This varies from a low of about 0.5% for very good coating to as high as 20% for poor coating; sometimes under exceptional conditions, it may run as high as 50%, but this is rare indeed. The only way in which an estimate can be made is by the application of experience, by comparison with tests run or systems installed on coatings known to be similar, or by field tests. The field work is similar to that described in Chapter 4, but the calculations may be simplified somewhat.

One method, illustrated in Figure 7-1, is to survey along the line after a polarization run of three or four hours, and locate the two points whose pipe-to-soil potentials are 1.0 volt and 0.8 volt. The section of line lying between these two points will then be at an average potential of about 0.9 volt, which is approximately the average of a line protected with distributed magnesium anodes. If, then, the line current is measured at each of these two points using the methods described in Chapter 2, the difference between these two line currents will be the total

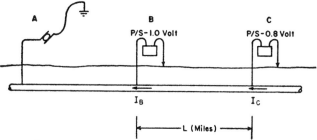

Figure 7-1. Determination of current requirements. Current is drained from the line at A, by a welding machine or other source of direct current. Points B and C, having *P/S* potentials of 1.0 volt and 0.8 volt respectively, are located; the average *P/S* between them is 0.9 volt. The current required to bring *L* miles of line to 0.9 volt is then $1_B - 1_C$, and the average current required for the line is

$$\frac{1_B - 1_C}{L} \quad \text{in amperes per mile.}$$

amount of current picked up on the line in the section. This quantity, divided by the length of the section, will then give the current requirements of the line in amperes per mile. Naturally, this figure can only be used for an entire line if the section chosen is truly representative; judgment must be used at this point.

If it is not possible or convenient to find the points which have the exact values mentioned, then readings can be taken which approximate these values, and both potential and line current can then be adjusted to find the desired quantity; this is illustrated in Figure 7-2; readings were taken at points A and B, and, by assuming a static potential of 0.6 volt and linear attenuation, it is determined that points C and D have the values of 1.0 and 0.8; since line current varies according to the same laws as potential, it is possible to determine the current values at the two latter points, and so to arrive at the desired quantity.

It is not to be expected that the estimate thus made will be highly accurate; it is intended to be merely preliminary.

Installation Procedure

After determining the approximate amount of current to be drained, the next step is to proceed with the installation. If, for example, the esti-

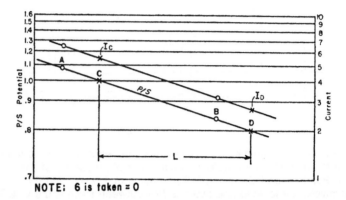

NOTE: 6 is taken = 0

Figure 7-2. Determination of current requirements. Measurements of *P/S* potential and of line current are taken at A and B; it is then assumed (in the absence of other information) that the static potential of the line was originally 0.6 volt. On this basis, the *P/S* and current curves are drawn as straight lines on semilogarithmic paper, and the points whose *P/S* are 1.0 and 0.8 are located, at C and D. The difference between the current values for these two points then gives the current demand of the line for the distance *L*, as in the preceding figure.

mated current demand is 0.5 ampere per mile, a site should be chosen within about a mile of the end of the section. A low, wet, spot should be picked, with relatively low soil resistivity. The number and size of anodes to be installed will be governed by the resistivity, and by the response of the line itself. As a rule of thumb, 50-pound anodes should be used only if the resistivity is 400 ohm-cm or lower; 32-pound up to 700 ohm-cm; and 17-pound above, with the possibility that full current output may not be had above 1500 ohm-cm.

The connection to the pipe should be made before more than one anode is installed; it will then be possible to observe the current output of successive anodes as they are connected, and installation should be halted before the average output per anode falls below 150% of the designed value. (For 10-year life, the design values for the three sizes mentioned are 285, 185, and 92 milliamperes, respectively.)

The site for the second group should be selected at a distance about twice that which the design current of the first group should cover. This process is then continued, installing groups at about double the design spacing, until the entire line is covered. Shunts should be installed, but no current limiting resistors; the connection between the shunt and the anode lead may be made with a temporary mechanical connection at this time. The mechanical arrangement of the anode station is subject to much variation in design; the one shown in Figure 7-3 is only one possibility.

Polarization and Final Adjustment

The anodes thus installed should be permitted to operate unrestricted for a period of three weeks or more. This will permit adequate polarization and stabilization of current output. After this time, a current output and pipe-to-soil potential survey should be made. Resistors should be installed where needed (see Figure 7-4), and the current reduced to the designed value. It is particularly important to check the potential at the

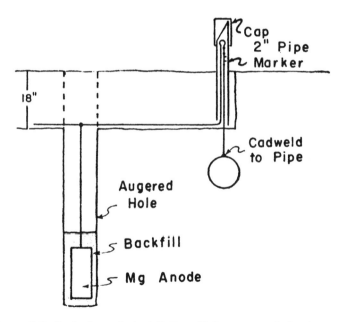

Figure 7-3. Typical anode installation. Only one anode is shown; the others may lie on a line parallel to or across the pipe line. Details of the assembly within the pipe maker are shown in Figure 7-4.

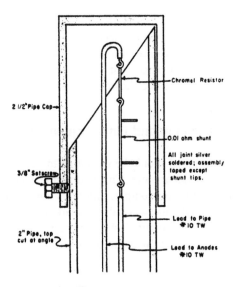

Figure 7-4. Anode station connection details. After final adjustment, the entire assembly within the pipe is taped up with electrical tape, leaving only the tips of the shunt exposed. If desired, the pipe may be poured full of asphalt to the lip, the wires first being pulled up above this level. The use of mechanical connections instead of silver-soldered joints is not recommended; the driving voltage is so low that a very little corrosion can effectively break the circuit.

midpoints between stations (if they are unequal in size, then at the low point). If these potentials should all be found above 0.85, then the installation is complete—and only one-half the estimated quantity of magnesium has been used.

Usually, however, this will not be the case, and more anodes will have to be added. Some can be added to existing stations if the current output is still high enough to permit it; otherwise, new stations will have to be installed between existing groups. This will in turn introduce new midpoints, which will have to be checked. Ultimately, the entire line will be brought under protection, with heavy installations only where they are needed, and thus with a near maximum of economy.

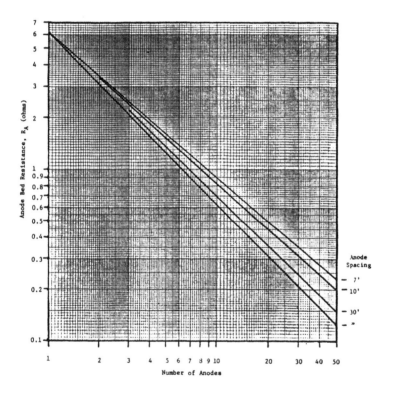

Figure 7-5. Anode bed resistance vs. number of anodes, 17-lb packaged magnesium anodes.

Chart based on 17-lb. magnesium anodes installed in 1000 ohm-cm soil in groups of
10 spaced on 10-ft. centers.

For other conditions multiply number of anodes by the following multiplying factors:

For soil resistivity: $MF = \dfrac{\rho}{1000}$ For 9-lb. anodes: $MF = 1.25$

For conventional magnesium: $MF = 1.3$ For 32-lb. anodes: $MF = 0.9$

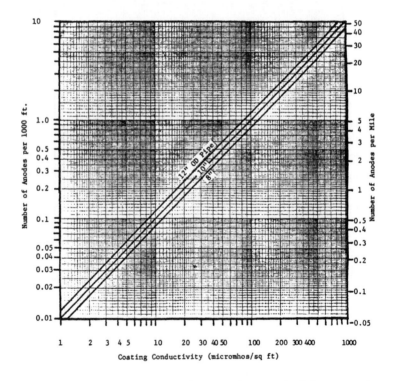

Figure 7-6. Number of magnesium anodes required for protection, coated pipe line.

Modification of Process with Experience

The procedure described is intended for the guidance of those with little or no experience in the technique. The experienced corrosion engineer actually combines the preliminary with the final phase and can "feel" his way along a strange line, observing the behavior of the current and of the potentials, occasionally backtracking to reinforce a group or to insert a new one, but generally moving forward until, by the time he has reached the end of the line, the installation is virtually complete. The final adjustment then becomes merely an adjustment and survey, with no new installation work required. The basic principle involved remains the same, however; instead of an extremely detailed survey, followed by an office design, there is a preliminary estimate, followed by an installation which is cut to fit as it is put in.

Sample Problems

1. What is the anode resistance of a packaged magnesium anode installation consisting of nine 32-pound anodes spaced 7-feet apart in soil of 2000-ohm-cm resistivity? Answer: from Figure 7-5,

$$1.0 \times \frac{2000}{1000} \times 0.9 = 1.8 \text{ ohms}$$

2. A coated pipe line has a coating conductivity of 200 micromhos per sq ft and is 10,000 feet long, with a nominal diameter of 10 inches. How many 17-pound magnesium anodes are necessary to protect 1000 feet according to Figure 7-6? Answer: Twenty 17-pound anodes will be necessary for the entire pipe line.

Summary

Cathodic protection systems have thus been designed using magnesium and zinc anodes to give cathodic protection to certain pipe lines when the economic considerations warrant it. Magnesium anodes are usually the first choice, but, since zinc anodes cost less and are more efficient, they may be used to advantage in certain long-term situations.

8

Hot Spot Protection

What "Hot Spots" Are

"Hot spots" have been defined in Chapter 1 as portions of uncoated lines most likely to cause corrosion problems. The question arises as to why there are uncoated lines at all. For the past 100 years, coatings have been available for pipe lines. During this time, techniques and materials have varied, but it is safe to say that no major pipe line has been built since 1900 that has not been coated in some way. The major method used since 1916 has been the hot applied coal tar plus wrapping system or the equivalent asphalt system. In the 1940s a cement-asphalt coating named "Somastic" provided excellent corrosion protection. During the 1950s, these alternatives were augmented by various tape systems which were easier to apply and almost as good. The newest development is the "thin-film" epoxy, which is shop-applied and essentially holiday free. The question, then, is why are there bare lines around? The answer is that flow and gathering lines were sometimes installed by marginal operators, and, consequently, there still are many buried uncoated lines throughout the United States. Therefore, the following material on handling bare lines should be useful.

Hot Spot Protection

Often the cost of complete cathodic protection for a bare or poorly coated line will prove to be prohibitively expensive. This is particularly likely to be the case where the line is of a semipermanent nature, or where its operation is intermittent so that small loss is occasioned by shutdowns, or where the land through which the line passes is such that damage claims are apt to be low or nonexistent. In any of these or similar cases, consideration should be given to the possibility of applying protection only to those sections of the line which are subject to the most aggressive attack; usually, by concentrating on these "hot spots," the leak rate can be cut to 5–10% of its original value, at a cost of perhaps 15% that of full protection.

This type of protective system, requiring small amounts of current to be drained from the line at different locations, leads almost automatically to the choice of magnesium anodes as the current source. The problem, then, is the selection from among the variety of anode sizes available, the determination of the locations where anodes are needed, and how many are to be installed at each location. Since the entire operation is based on securing an adequate degree of protection at a minimum of total expense, there can be no large expenditure on elaborate surveys and design; on the other hand, the promiscuous installation of anodes here and there on the line is certain to be wasteful.

Locating Hot Spots

There are three basic methods for locating the hot spots: (1) where there have been leaks, or where there is visible evidence of corrosive attack, there are hot spots; (2) surface potential surveys can disclose the presence of current flow which is evidence of active corrosion; and (3) soil resistivity surveys will indicate the locations at which corrosion is likely.

On lines which have already started to give trouble, it would indeed be foolish to ignore the evidence of hot spots which is afforded by the leaks which have already occurred. Similarly, when some protection has been installed, the occurrence of leaks indicates that more is needed at that particular spot. Visual inspection of the line when an excavation has been made for any purpose will often make it possible to locate areas where corrosion is progressing but has not yet developed into an actual leak.

In the absence of such direct evidence of corrosiveness, surface potential surveys may show the points at which current is leaving the line;

these will be the anodic, or corroding, areas. The method of making such a survey was described in Chapter 2. For the present purpose, the electrode spacing should be about 25 feet in general. When a "reversal" is discovered, the spacing may be reduced to a much smaller value; often the hot spot can be outlined with an accuracy of a foot or two. If the sole purpose is the location of hot spots, there is no need for referring the potentials to the pipe itself; all that is needed is the evidence of current flow in the soil, as indicated by differences between two copper sulfate electrodes. For a more positive indication, when evidence of an anodic spot is found, readings should be taken with one electrode over the line and the other 5 or 10 feet away at right angles; both sides should be used, to eliminate the effects of any current flow across the line from extraneous sources.

Figure 8-1 shows the current flow along the line and in the soil in the vicinity of an anodic area, and also the potential profile which would be obtained at the surface. When the point of reversal is located, the lateral

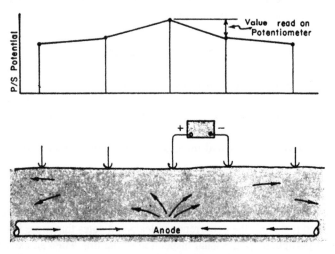

Figure 8-1. Hot spot location by surface potentials. In the vicinity of an anodic area, or hot spot, the current in the pipe is flowing *to* the anode, while in the soil it is flowing *from* the anode. Thus the potential drop along the surface above the pipe is away from the hot spot, as shown in the diagram. The point of highest potential is the center of the anodic activity.

readings should be taken; if they, too, indicate flow in the soil away from the point, there is unquestionable evidence of a corroding spot.

Almost no hot spot surveys are made by the methods just described because of the expense involved. The method is very precise, however, and is sometimes useful in congested areas, or when dealing with high unit cost lines, such as pipe-type high-voltage cables.

The soil resistivity survey is somewhat less exact, but much less expensive, and is the method used for the vast majority of such surveys. Readings may be taken in either of two ways: with the four-terminal method (most common) or with a single rod, in which case the ideal time to make the survey is during construction, with the readings being taken in the bottom of the ditch. Spacing of readings, and methods of plotting, have already been described in Chapter 1; there remains the task of locating the hot spots.

These are regions of *relatively* low resistivity, and it is not easy to formulate precise rules for their selection. As a general guide, it may be taken that soils below 1000 ohm-cm are almost always corrosive, except when they are the high ''peaks'' along a line of generally much lower resistivity. Soils above 10,000 ohm-cm are generally considered to be noncorrosive, except when they appear as narrow valleys on the profile between stretches of soil of higher resistivities. But even in these cases, they are never as rapidly corrosive as are the low-resistivity soils. In-between the values of 1000 and 10,000, soils must be judged by comparison with their neighbors; the relatively low sections will be found to be the site of the major attack.

Usually, a company or a group of companies operating in a general area will adopt a standard procedure for the selection of hot spots and their protection. This is a logical step since all of the gathering lines in a given field, for example, can be expected to have about the same projected useful life. In one field, where the mean resistivity is about 1500 ohm-cm, with a minimum in the neighborhood of 300 and with hardly any spots over 20,000, the following procedure is almost standard for the protection of bare gathering lines:

1. All lines in soil below 1000 ohm-cm to receive protection at the rate of 2 milliamperes per square foot.

2. All lines in soil between 1000 and 1500 ohm-cm to receive 1½ milliamperes per square foot, unless the soil on *both* sides of the section are of lower resistivity, in which case no protection is to be applied.
3. Lines in soil between 1500 and 3000 ohm-cm to receive 1 milliampere per square foot, but only when the soil on both sides is of higher resistivity.

In the application of these rules, the first step is to go through the plotted resistivity profiles and mark, usually along the bottom of the sheet, the lengths which are found to be hot spots, by these definitions. Three colors may be used, or lines of different weight, to distinguish the three classes. This locates the hot spots.

Anode Selection and Spacing

The amount of current which an anode will deliver when connected to a bare line is determined by the size and shape of the anode, the composition of the alloy used, the potential to which the pipe is brought by the protection, and the resistivity of the soil. The life of an anode depends upon the current output and the size of the anode (assuming that all anodes operate with the same efficiency, which is not too great an assumption to make).

With all of these variables to contend with, the selection and spacing of anodes along the hot spots already determined is not an easy matter. However, with an adopted set of design principles, the process can be systematized to the point that it can be done almost mechanically.

The first decision to be made is the intended life of the installation; in the majority of cases, 10 years is the design figure used. To attempt to design for longer periods with magnesium anodes is not economical, because of the reduced efficiency of magnesium at low current output. To design for shorter periods is not usually economical, except in the obvious case of piping whose useful life is considered to be less than 10 years, in which case the design life should be made equal to the useful life of the system. Perhaps it should be noted that it is not at all rare for piping whose original anticipated life is short to undergo successive extensions of usefulness; the "temporary" line still in use after 20 years is fairly common.

The next thing to determine is a standard set of anodes to use. The simplest set is, of course, only one size. But for maximum effectiveness in fitting each anode to the specific situation, a more elaborate set is needed. Anodes of standard alloy are available in sizes from 2 or 3 pounds up to 200 pounds. For hot spot protection, it is usually sufficient to use anodes of three standard sizes: 17, 32, and 50 pounds.

When the soil resistivity is very low so that even a 50-pound anode puts out so much current that its life is less than 10 years, two or even three anodes can be used in a single hole, one on top of the other, and the current will be reduced (per anode) and the life correspondingly extended.

At the other end of the scale, in soil of high resistivity, the small (17-pound) anode does not put out enough current, so we add to our arsenal a 17-pound anode of high-purity magnesium. Commonly known as a "high-potential" anode, this will extend the useful range a bit. At still higher resistivities, we can introduce a long slender anode of the same high-potential alloy. Because of its greater area, it will have a still higher current output.

In most instances, this set will be all we will need. If we did need to operate in still higher-resistivity soils, and still get current outputs consistent with 10-year life, we would first go to lengths of extruded magnesium rod, and then to extruded magnesium ribbon. However, both of these function at 10-year life in soils of higher resistivity that are usually considered corrosive, as far as hot spot protection measures are concerned. Summarizing, our arsenal of weapons comprises the following distinct anodes:

3 50-lb. standard alloy . . . . 350 ohm-cm
2 50-lb. standard alloy . . . . 370 ohm-cm
1 50-lb. standard alloy . . . . 550 ohm-cm
1 32-lb. standard alloy . . . . 780 ohm-cm
1 17-lb. standard alloy . . . . 1500 ohm-cm
1 17-lb. high potential alloy . . . . 1900 ohm-cm
1 17-lb. high-potential alloy, long . . . . 2900 ohm-cm

For each anode listed, there has also been given the soil resistivity in which it will deliver enough current (when connected to a steel pipe polarized to 0.85 volt) to project an anode life of 10 years. The significance of these values is that each anode should be used in soil of resistivity equal to or higher than the values shown, in order to obtain a life of 10 years or more. If there are extensive areas of resistivity lower than 350 ohm-cm, special treatment is necessary—either larger anodes (special) or, more economically, an installation aimed at a shorter life.

The spacing of anodes along the line is determined by the current output of the size selected in the soil at each section, and by the diameter of the line, so that each section receives the intended current density. Tables can be prepared, showing for each anode size the current to be expected in various soils, as well as the projected life corresponding to each current, and the spacing for various sizes of line which will then supply any desired value of current density. Such a table is very complex because of the large number of variables and does not lend itself well to design use where several hundred values have to be read out in a limited amount of time.

For actual design use, an abridged table should be made, showing only the anodes whose use is contemplated. Furthermore, each anode should be shown only with those values of resistivity in which it is useful, according to the particular policies being followed (current density, desired life) and the spacing for the pipe sizes under study. The following table is an example. It is designed around the set of anodes already named; is based on a minimum life of 10 years; and is limited to 2 milliamperes per square foot. All are on bare pipe of three diameters: 2 inch, 3 inch, and 4 inch.

It must be understood that Table 8-1 applies only to *one* set of conditions: 10-year minimum life, current density of 2 milliamperes per square foot, the specific set of anodes listed, and the three pipe sizes given. It is also implicit that the current furnished will be just enough to bring the pipe to 0.85 volt with respect to a saturated copper sulfate electrode. This is an approximation, of course, and it is one of the assumptions which may not be fully realized, and thus makes the design perform with less than perfect agreement with theory.

Another difficulty lies in the matching of the anodes with the plotted resistivity. Any one hot spot will show several values of this variable, and the points of actual reading will not coincide with, and will usually be less numerous than, the points of anode installation. Hence, the actual resistivity in which a certain anode is placed is not precisely known, but must be inferred from the adjacent readings. In practice, this has not been found to be an insuperable difficulty; errors are intro-

Table 8-1
Anode Spacing Vs. Soil Resistivity

Soil Resistivity ohm-cm	Anodes	Current Milliamp	Life Years	Density Milliamp Sq. Ft.	Spacing in Feet Pipe Size		
					2"	3"	4"
360	3 50S	840	10.3	2	675	456	354
380	2 50S	569	10.1	2	457	310	240
400	"	541	10.6	2	435	295	228
420	"	516	11.2	2	415	281	218
440	"	493	11.7	2	396	268	208
460	"	472	12.2	2	379	258	199
480	"	453	12.7	2	364	247	192
500	"	435	13.2	2	350	237	184
520	"	419	13.7	2	336	228	177
540	"	404	14.3	2	324	220	170
560	1 50S	285	10.1	2	229	155	120
580	"	275	10.5	2	221	150	116
600	"	266	10.8	2	214	145	112
620	"	258	11.1	2	207	141	109
640	"	251	11.5	2	201	136	105
660	"	243	11.8	2	195	132	102
680	"	236	12.2	2	189	128	99
700	"	229	12.6	2	184	125	97
720	"	223	12.9	2	179	122	94
740	"	217	13.2	2	174	118	92
760	"	211	13.6	2	169	115	89
780	1 32S	186	10.0	2	148	98	78
800	"	180	10.2	2	144	95	76
820	"	176	10.5	2	141	93	74
840	"	172	10.8	2	138	91	73
860	"	168	11.0	2	135	89	71
880	"	164	11.3	2	132	87	69
900	"	161	11.5	2	129	85	68
920	"	158	11.8	2	127	83	67
940	"	154	12.0	2	124	81	65
960	"	151	12.3	2	121	80	64
980	"	148	12.5	2	119	78	63
1000	"	145	12.8	2	116	76	61
1200	"	121	15.3	2	97	64	51
1400	"	104	17.8	2	83	55	44
1600	1 17S	91	10.7	2	73	48	38
1800	"	81	12.1	2	65	43	34

Table 8-1 (Cont.)

Soil Resistivity ohm-cm	Anodes	Current Milliamp	Life Years	Density Milliamp Sq. Ft.	Spacing in Feet Pipe Size		
					2″	3″	4″
2000	1 17H	91	10.8	2	73	49	38
2200	″	82	11.9	2	66	45	35
2400	″	76	12.9	2	61	41	32
2600	″	70	14.0	2	56	38	30
2800	″	65	15.1	2	52	35	27
3000	1 17LH	93	10.5	2	75	51	40
3200	″	88	11.2	2	71	48	37
3400	″	83	11.8	2	66	45	35
3600	″	78	12.5	2	63	42	33
3800	″	74	13.2	2	60	40	31
4000	″	70	13.9	2	57	38	30
4200	″	67	14.6	2	54	36	28
4400	″	64	15.3	2	51	35	27
4600	″	61	16.0	2	49	33	26
4800	″	59	16.7	2	47	32	24
5000	″	56	17.4	2	45	30	24

duced, but they are not serious; certainly the technique works in a satisfactory manner.

More general tables can be made in which the output of each size of anode is given for a wide range of soils complete with the projected life, spacing for various pipe sizes at different current densities, and other data. Table 8-1 is the result of splicing together the pertinent parts of a set of such tables. For a different set of conditions, a different set of parts would be taken and spliced together.

Field Installation

When the data from such a study is taken into the field for the actual installation, the resistivity measuring instrument should not be left behind. Most resistivity surveys are made by pacing rather than chaining, and the exact relocation of points where readings were taken may not be easy. Besides, when two readings of 780 and 800 ohm-cm have been taken 400-feet apart, it is never possible to be sure that the soil between them is in the same range; it might turn out to be 3000. This can affect profoundly the actual installation. Actual resistivity in which anodes are installed should always be verified by measurement at the time of installation; an alternative method is described here.

Field Design

Another way out of the dilemma described is to make the actual design in the field. To do this, it is necessary to have one person doing the computing while another does the measurement. Usually, a three-man crew will be needed because of the necessary staking. In this procedure, the crew proceeds down the line, with resistivity measurements being taken as needed. The design is worked out (from the specially prepared table, as just described) and the resistivity verified at the actual locations of most anodes, which are marked with stakes as the decisions are made. In long stretches of reasonably uniform soil, with closely spaced anodes, it is not necessary to check every single site. With practice, this method will be found to be very effective and economical.

Zinc Anodes in Hot Spot Protection

Because of their lower driving voltage, and consequent lower current in a given application, zinc anodes find a place in hot spot protection of bare lines only in soils of the lowest resistivity. The array of anodes just given can be extended for use in lower soils by the addition of one or two sizes of zinc anodes. Since these are available in such a large variety of sizes, as well as several compositions, no tables have been computed for these.

As mentioned earlier, zinc differs from magnesium in that the current efficiency is good even at very low values of current drainage. This makes it more adaptable for very long-life installations—up to 30 years. Ordinary gathering systems would not justify such installations, even with anticipated useful life this long; repeated installations at 10-year intervals would be cheaper. This is not the case, however, with many distribution systems, where pavements, narrow easements, and generally difficult access make very long-life installations much more attractive. In these situations zinc is readily applicable.

Installation Details

In general, anodes should be set about 10 feet from the line and at a depth of eight feet. This depth may be reduced if permanent moisture is present at shallower depths and should be reduced if the lower strata are found to be of high-resistivity soil. Placing the anodes closer to the line will result in a slight increase in current output (up to perhaps 50% at 2-foot spacing), but the additional current so obtained nearly all flows to the pipe immediately adjacent, so the gain is more apparent than real. Sometimes, of course, the presence of other lines in the same right-of-way makes closer spacing necessary; in this case, the needed separation can be obtained by increased depth.

For most of the conditions encountered, each anode will be connected to the line by its own individual lead wire. Shunts and current-controlling resistors may be installed, though on hot spot projects the shunts are frequently omitted. If the resistor is to be adjusted at the same time the anode is installed, it will be necessary to make an allowance for the expected reduction of current output which will come with polarization and with the installation of additional anodes. Some experience is necessary to make this allowance, and it can never be made with precision; perhaps 40–50% is a conservative figure. Better results may of course be had if the final adjustment is deferred until all of the anodes have been in operation for a few weeks, but this procedure is obviously more expensive; in the absence of considerable experience, however, the added expense is usually justified.

Under conditions which make the attachment of lead wires to the line particularly difficult, such as will be encountered in swampy areas, a collector wire may be laid parallel to the pipe, and the anodes connected to it. The collector wire may then be connected to the pipe at convenient locations; these should not be more than 400- or 500-feet apart, and should be even more closely spaced, except under very difficult conditions.

It may or may not be possible to leave some kind of permanent marker to indicate the location of each anode; often this cannot be done, but steps should be taken, by careful measurement and recording, to make sure that the locations are accurately known and are recoverable for testing purposes.

Supervision and Control

An installation, made in accordance with the procedures outlined here, cannot be considered as perfect or complete. As mentioned earlier, the method described is intended to afford a reasonable amount of protection; it is quite likely that some hot spots will be missed, either because they are very short and were skipped over in the survey, or because they are due to highly local conditions not observable at the surface of the earth. From time to time, then, these may make themselves known by leaks. There may also be leaks in spots where the protection was not quite adequate. In either case, additional anodes should be installed.

It is a recommended practice on bare lines, even where no hot spot system has been installed, to install one or two anodes at the site of every leak known or suspected to be due to corrosion, or at every place where any excavation of the line for any reason shows evidence of attack or exposes corrosive soil. Anodes installed in this way are very inexpensive and are to be considered as a low cost form of leak insurance. The repair crews can easily be instructed in the method of installation, and the losses involved by the placing of unneeded anodes will not be as great as the losses avoided by those which are useful.

Since hot spot protection is essentially low-cost protection, there is as little point to elaborate resurveys and inspections as there is to elaborate preliminary surveys. However, at two- or three-year intervals, a check should be made on most of the anodes to see how the current output is holding up. Those in particularly difficult spots may be skipped, provided there is no reason to suspect damage to lead wires, and provided not too many are skipped in one stretch. A complete survey at regular two-year intervals is not too expensive and will probably justify itself in continued protection.

Summary

The above-mentioned procedure is an example of how the difficult problem of cathodic protection of bare pipe lines in soil may be solved. If a cathodic protection engineer was involved in the initial installation of these lines, there will be no "hot spot" problems. The lines will already be coated.

9

Stray-Current Electrolysis

Stray Current Corrosion

Cathodic protection was an offshoot of the studies on interference corrosion caused by the many electric railways and trolley lines which spread over the United States in the early part of this century. The direct current discharged into the ground caused many failures of nearby piping systems. "Electrolysis committees" were formed to find solutions to the corrosion problems caused by the "stray currents," and much of our knowledge of corrosion current measurement and bonding methods was developed long before 1913, when R. E. Kuhn started cathodic protection in New Orleans.

Today, streetcars are almost all gone, except in some metropolitan areas, and diesel engines have replaced the electric trains once used on the railways. However, there are several experimental direct-current transmission systems now proposed for transcontinental networks, and the stray-current problem may arise again.

Sources of Stray Currents

Some of the electrical currents which corrode pipe lines are those which arise from galvanic potential differences between various parts of the structure in contact with the earth. Others, however, are the result of the leakage of current from some electrical system so that part of the current path is through the earth. Whenever a pipe line lies within such a current path, there is an opportunity for current to enter and leave the line; at the points where it leaves, it will corrode the pipe. Because of the inherently accidental or unintentional nature of such currents, they are usually known as *stray currents*, and the damage they do is known as *stray-current electrolysis*.

By far the greatest source of stray current is the electric railway, or its urban counterpart, the street car. Figure 9-1 shows in simplified form the route taken by the current. It should be noted that the rail is supposed to provide the return path; it might be assumed that, if the rail joints were adequately bonded, no trouble would result. This, however, is not true. The electric current does *not* "follow the path of least resistance." At least it does not all follow that path. Electric current, when offered two or more parallel paths, divides itself between them inversely as the resistance. If the rail path has one-tenth the resistance of the earth path, then it will carry 10 times as much current—but the remaining portion, the one-eleventh of the total current which flows through the earth, is sufficient to do a great deal of damage.

The situation is not always as simple as that shown in the diagram. Sometimes the affected line may lie at a considerable distance from the tracks; perhaps the point of attack is near the crossing of two lines,

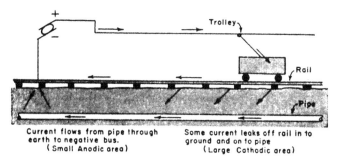

Current flows from pipe through
earth to negative bus.
(Small Anodic area)

Some current leaks off rail in to
ground and on to pipe
(Large Cathodic area)

Figure 9-1. Stray-current electrolysis. Return current from streetcar divides, part going back to substation along rails, and part leaking off rails onto pipe line. Near the substation this current flows from the pipe line through the soil to the rail system, causing corrosion of the pipe. Installation of a metallic bond from the pipe to the negative bus at the substation will avert the damage.

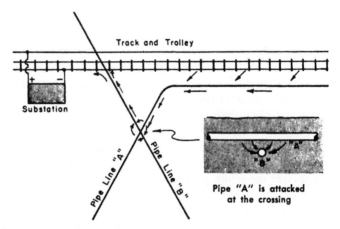

Figure 9-2. Complex stray-current exposure. The point at which damage occurs to pipe line "A" is remote from the offending street railway tracks. Such a condition may be detected by fluctuating potentials or currents in pipe line "A" and located by a line current survey.

where some current leaves one line and enters the other. Figure 9-2 shows a case which proved to be very difficult to detect, in view of the distance from the nearest track to the point of damage; yet the relation is clear, once it has been traced out and diagrammed.

There are other possible sources of stray current in addition to electric railways. Almost any DC power network is capable of causing damage in this way, although most of them are innocent. Mine railways, cranes, and other machinery using DC should be suspected. Frequently, there are severe exposures in and near chemical plants using electrolytic processes. Welding equipment, particularly when employed in production work, is a common source of trouble, although rarely is the damage done at any great distance from the equipment; this makes it much easier to locate.

Finally, it is unfortunately true that the direct currents introduced deliberately into the earth for the purpose of applying cathodic protection to one structure, are capable of doing great damage to other structures which occupy the same earth. This is not strictly "stray current"; the damaging effect is altogether accidental, but the presence of the current in the earth is not accidental. Consideration of this problem is deferred until later.

Detection of Stray Current

Whenever the measurement of any of the electrical quantities connected with a pipe line—line current, pipe-to-soil potential, or any other—shows fluctuating values, there is a case of stray current at hand. Often it is very helpful to leave the instrument connected and just watch the fluctuations for a while. The rapidity and nature of the changes will frequently give a clue as to the origin, and it may be that the actual operation of the offending system can be observed; for example, the change when a street car passes may unquestionably locate the trouble.

When location of the source is not possible by this means, a recording meter should be connected. Even when the value recorded is not the true value, because of lower sensitivity of the instrument, the record over 24 hours can be very useful. For example, if there is a cessation of fluctuation for the noon hour, then the source is industrial machinery rather than transportation equipment. By the application of similar reasoning to records of several successive days, it is usually possible to track down the DC system from which the stray current is coming.

Remedial Measures

Good coating is effective in reducing the *total* damage done in a stray-current exposure; but a line with good coating may experience more rapid penetration rates at the few holidays present than would be incurred on a bare line. Coating alone cannot be depended upon as protection from stray-current damage, just as it is not enough for general soil corrosion.

The installation of insulated joints on either side of an electric railway will do much to alleviate the attack; the exposure will be limited to the short section between the joints, and this can be handled by the other measures enumerated here. Such joints, if not properly installed, can actually act to concentrate the attack without appreciably minimizing it.

The direct approach to stray-current problems is the installation of bonds between the threatened pipe and the negative terminal of the DC system which is doing the damage; this affords a low-resistance path by which the current, collected on the line over long sections, may leave without doing harm in a small, concentrated section. "Solid" bonding, where the lowest possible resistance path is afforded, is the simplest method, and is often effective; bonding with specific resistance values, however, is often necessary.

In some locations, where the amount and even the direction of the stray-current flow depend on the position of the cars or trains, it has proved practical to install relay-controlled bonding switches, usually referred to as electrolysis switches, which operate in such a way as to interrupt the circuit or close it, depending on the magnitude and direction of the potentials involved. A modification of this system uses rectifier units as electrical check valves, permitting current flow from the pipe line when the potential is in that direction, but blocking flow of current to the line when the potential reverses.

Negative Bus Bonding

In a situation such as that illustrated in Figure 9-1, the damage can be eliminated by the installation of a metallic connection or bond between the pipe and the negative bus at the substation. When the geometry is as simple as in the illustration, the "area of maximum exposure" will be in the immediate neighborhood of the substation. In more complex cases, a survey will be required to locate this area.

The installation of such a bond affords a measure of cathodic protection to the line. In an extensive network, involving many pipe lines and many tracks, such as would be found in the gas system of a city with street cars, almost the entire gas distribution system may be placed under cathodic protection by the installation of such bonds. Does this involve expense to the street car company? Since they are furnishing the power, the first answer which occurs is in the affirmative. But consider that the installation of bonds actually *decreases* the over-all loop resistance of their circuit, and it is seen that actually less power is required to perform a certain amount of work. Yet it does actually cost the street railway company—not in power, but in increased damage to its rail system. The presence of the bonds means that a higher percentage of their return current leaks off the rails and returns by the earth-pipe-bond path than would otherwise be the case. It is for this reason that efforts are sometimes made to install resistance bonds which will limit the current to that which is required to prevent damage to the lines, without actually affording them cathodic protection against their normally occurring soil exposure.

Exposure Areas

If the stray-current effect on a pipe line were steady, the areas of maximum exposure could be readily located by either a line current survey or a surface potential survey to determine the regions in which current is leaving the line. The fluctuations make the problem somewhat more complicated, and special measures are required.

One approach is a modification of the line current survey. Current measurements are made simultaneously at two locations, some 20–40 pairs of readings being taken. Corresponding pairs are then plotted on a graph, as shown in Figure 9-3; if the straight line formed is at a 45°

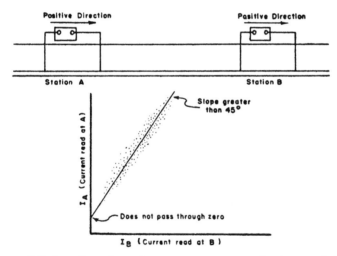

Figure 9-3. Location of maximum exposure area. In the graph illustrated the slope greater than 45° being closer to the axis of I_A indicates that the stray-current component at A is greater than that at B; hence, there is current loss in the section AB; the intersection above the axis indicates a small steady component at A; that is, there is some current loss in the section due to sources other than the stray-current exposure. If the points fail to define a straight line, there are two or more sources of stray current present. A similar technique can be applied to surface potential surveys.

angle and passes through the origin, then there is no current gain or loss in the section involved; if it is at an angle other than 45°, then there is current gain or loss due to the stray-current exposure; if it fails to pass through the origin, there is a steady gain or loss, independent of the fluctuating currents. Finally, if the points do not fall somewhat near a straight line, there is a stray-current source affecting one of the points and not the other.

As soon as the set of readings for one pair of joints is completed, one crew is leap-frogged ahead and another section studied. If there are more than two crews and sets of instruments available, more than one section can be studied at one time. The readings may be synchronized either by use of signalling equipment (a pipe locater transmitter and receiver can be used) or by the use of synchronized watches. Such surveys seldom cover very great lengths of line, usually being limited to areas which appear to be likely zones of exposure from a study of maps.

Potential Surveys

Although the study of line currents, as just described, yields the most positive and definite evidence of actual current gains and losses—and these are the direct correlates of corrosion—much information can be gained by a study of pipe-to-soil potentials, and of pipe-to-pipe, or rather, of structure-to-structure potentials in areas where trouble is suspected. To do this effectively, it is necessary to measure a number of potentials simultaneously—and it is important that this be exact, for a difference of less than a second could disguise or distort a relationship beyond recognition.

Technical Committee T-4B of the National Association of Corrosion Engineers first described a device (*Corrosion*, Vol. 13, p 799t, December, 1957; also NACE publication no. 57-26) which performs this difficult task quite effectively. A number of capacitors are connected to the various terminals between which potentials are to be measured. When a set of readings is necessary, a button is pressed (this can be done remotely) and all capacitors are simultaneously disconnected, thereby "freezing" their separate potentials. These are then read, one at a time, with a vacuum-tube voltmeter.

The instrument described in the reference had 19 circuits, being designed primarily for the investigation of cable sheath corrosion; rarely would so many be needed in pipe line work, and a very simple modification of the device, using two or three 4-mf capacitors, can readily be improvised. With such a small number, the relays and remote control devices can be dispensed with, and even the rotary switch is unnecessary; clips can be shifted from one terminal to another as needed. The important feature of the whole device is that the capacitors must all be disconnected from their potentials at the same time.

Secondary Exposure

Figure 9-2 shows a situation in which an exposure area lies at some distance from the tracks. Current leaks off the tracks onto pipe line "A," flows down it to the crossing, then leaks from "A" to "B" and follows "B" to the vicinity of the substation or track. To avert the damage, bonds must be installed at both places; the installation of either bond, alone, will aggravate the exposure at the other location by decreasing the resistance of the total return path. Such exposures, and others even more complicated, can only be found by very careful and thorough surveys; often they are disclosed by the occurrence of leaks. Surface potential surveys, however, will indicate fluctuating potentials; as long as these remain unexplained, there is an exposure somewhere on the line; careful work will eventually track it down. *Fluctuating potentials or line currents must never be ignored.* They often turn out to be harmless, but until their source is known, they should remain a case to be investigated.

On well-coated lines in high-resistivity soil, and more particularly in northern latitudes, wide fluctuations in potentials are often observed which can only be associated with natural earth currents. These are much more violent during periods of magnetic activity, as when the northern lights are active, but sometimes may be continuous for weeks. Apparently these do little if any harm; they usually disappear once a line has been placed under cathodic protection. However, they do make current requirement surveys quite difficult, as they virtually preclude the taking of any "static" potentials.

Summary

It is seen that the major problem causing stray-current electrolysis in the past was direct-current leakage from electric streetcars and trains.

However, a present cause is direct current from a foreign unbonded pipe line under cathodic protection. All the techniques we have learned to measure electrical current, including the recording meter, may be needed to determine the problem and its source. Insulating flanges and negative bus bonding are often used to minimize the damage caused by electrolysis.

10

Interference in Cathodic Protection

The Problem

In the preceding part stray current was defined as leakage current from some electrical system, flowing accidentally through the earth. If this is capable of causing damage, so also are the currents deliberately caused to flow through the earth in the application of cathodic protection. In the former case, it is a problem of the corrosion engineer to avert the damage to the underground structures under his care; in the latter, it is his responsibility to take steps to prevent damage to all other structures by the action of his cathodic protection systems. It must never be supposed that the current flow is confined entirely to the earth which actually lies between the anode and the protected structure. Figure 2-9 (Chapter 2) shows a cross section of the field surrounding a pipe line at a considerable distance from the anode. It will be noted that the field is symmetrical; that is, there is no visible or measurable difference in the amount of current which is arriving from the side on which the anode is placed and that which is arriving from the other.

Current flow can be detected at the surface of the earth by the use of two electrodes and a potentiometer, at a distance of several hundred feet on either side of a protected pipe line and at almost any distance from the anode. It can also be detected at a distance of a few thousand feet in any direction from the anode. There is no definable limit to the field; more sensitive instrumentation might extend the distances even further.

Wherever there is current flow in the earth, from whatever source, a piece of metal buried in that earth may function as a part of the current path, collecting current over a part of its surface and discharging it—with attendant corrosion—from another part. The amount of current so picked up and discharged is controlled by many factors: the coating, if any, on the structure; the length of the structure in the direction of current flow; the potential gradient or the current density at the point of exposure; and the ability of the structure to carry current. A bare pipe line, lying in the direction of current flow, close to an anode bed (where the current density and potential gradient are great) is in a very hazardous exposure; a well-coated line, remote from an anode bed, and traversing the current field at an angle, is much less seriously endangered.

The problem of cathodic interference is simpler in many respects than that of stray current. The exposure is in general steady rather than fluctuating, so that more accurate measurements and adjustments can be made. The source of the current involved is under control; the rectifier can be switched on and off as desired, or an interrupter may be used, so that the effects of the unit can be clearly identified and studied. Under the worst conditions, the exposure can be extremely severe, because of the magnitude of the currents involved; *all* of the current return is by the earth path, instead of merely that portion which leaks off, as with stray currents.

Basic Solutions

There are three fundamental approaches to the problem: (1) design which aims at minimizing exposure, (2) bonding to afford a metallic return of current collected by a foreign line, and (3) auxiliary drainage of the collected current.

Design

The current density in the earth is far greater in the vicinity of an anode bed than it is elsewhere; consequently, this is the area of most hazardous exposure. Every effort should be made to select sites for the installation of rectifiers which are remote from foreign lines. Figure 10-1 shows the current path in the case of a crossing foreign line which pass-

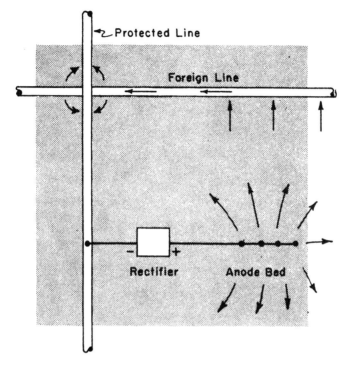

Figure 10-1. Interference. Some of the current flowing from the anode bed to the protected line collects on the foreign line, flows along it toward the crossing (from both sides), and then discharges through the soil to the protected line. Damage is inflicted on the foreign line in the neighborhood of the crossing.

es near the anode bed; the situation is similar, but much milder, when the crossing line is remote. Figure 10-2 shows a case of interference which is serious only when the ground bed is quite close. The major design feature which will minimize such damage is obviously that of keeping anode beds far away from foreign structures, insofar as this is possible.

Crossing Bonds

In a situation such as that illustrated in Figure 10-1, the remedy is the installation of a metallic connection, or bond, between the two struc-

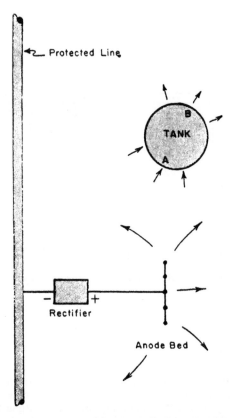

Figure 10-2. Interference (radial current flow). When a structure lies in a region of heavy current density, such as the tank shown close to the anode bed, it may pick up current at A and discharge it to earth at B, with resultant damage at the discharge area. Sometimes, but not often, an isolated metallic structure lying near a protected line can undergo the same kind of damage.

tures, at or near the point of crossing. Sometimes a "solid" bond is used; i.e., one without a resistor. This usually results in affording a considerable measure of cathodic protection to the foreign line, and if it is bare or has poorer coating than the protected line, jt may receive too large a share of the current. Clearly the fair thing to do is to drain across the bond just enough current to prevent inflicting any damage on the foreign line; to do so requires the use of a resistor in the bond.

The direct method of installing and adjusting such a bond is illustrated in Figure 10-3. A copper sulfate electrode is placed in the soil between the two lines at the point of crossing. This will usually involve an excavation, exposing a vertical wall of soil where the two lines are closest; sometimes a long electrode can be placed properly in a rodded hole. The connections to the two lines are made and the resistor adjusted until there is no change in the potential of the *foreign* line with respect to the electrode when the rectifier is switched on and off. It is helpful to measure the ''short-circuit'' current between the two lines first, using the ''zero-resistance ammeter'' circuit diagrammed in Figure 10-5; from this an idea may be gained of the size of conductor needed for the bond. The pipe connections may usually be made by the

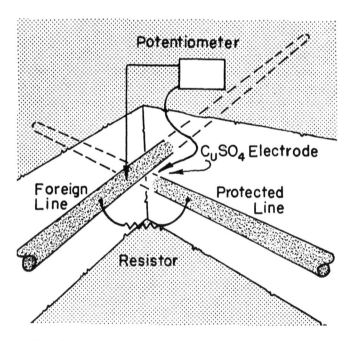

Figure 10-3. Adjustment of crossing bond. The copper sulfate electrode is placed between the two lines at the point of crossing. The resistor is then adjusted by trial and error until there is no change in the potential of the foreign line with respect to the electrode when the rectifier is turned on and off.

thermite process, and Chromel or Nichrome wire used for the resistor; the joints should be silver soldered. It is much easier to determine the length of resistor wire needed by trial and error than it is by calculation.

Calculation of Bond Resistance

Instead of the direct approach just described, a more refined method makes it possible to calculate the value for an interference bond. The method used is that of determining *circuit constants*. The complex circuit can be converted into a simple equivalent circuit which will duplicate the behavior of the two pipe lines at the points of interest, and hence can be used to solve the problem at hand. The method will be illustrated by considering first the simplest possible case.

Suppose there is a line section under protection by a single rectifier (or under test from a single drain point) (Figure 10-4). Suppose this line is to be crossed by a single foreign line outside the anode field. Further, to get the simplest possible case for illustration, suppose that there is no potential difference between the two lines at the point of crossing in the absence of the cathodic protection.

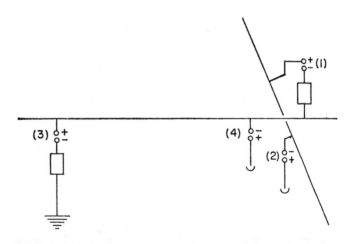

Figure 10-4. Schematic diagram for determining correct value of bond resistance and rectifier current to eliminate crossing interference.

The first step is to adjust the rectifier to a value which will give adequate protection to the line. Then, with the rectifier current being interrupted, a survey is made along the foreign line in the vicinity of the crossing until the point at which the pipe-to-soil potential of the foreign line is at a minimum. Normally, this will be at, or very near, the crossing. This point is carefully marked and is indicated as (2) in Figure 10-4. (The location refers to the precise position of the electrode; the actual contact to the pipe may be made at any convenient near-by point, if the pipe is continuous.) The point (1) is a pair of terminals or test leads connected to the two lines at or near the crossing; these are leads in which a bond may be installed if needed. The point (4) is a location for an electrode over the protected line; its exact location will be described later. Finally, the point (3) represents the terminals of an instrument (ammethe terminals of an instrument (ammeter) inserted in the rectifier circuit for the purpose of measuring its current output.

Always, in making interference tests of this kind, a sketch like Figure 10-3 should be prepared, showing everywhere readings are to be taken; the expected polarity should be indicated by + and − signs shown at each numbered pair of terminals. Then, when the test is under way, if the reading corresponds to the indicated polarity, it is recorded as +; if it is opposite, it is recorded as −.

The test procedure is as follows: the rectifier current, or temporary test current, is interrupted, and values of V_2 and V_1 are determined for both conditions (on and off). V_2 is the voltage across terminals (2); i.e., the P/S potential of the foreign line; V_1 is the voltage across terminals (1), or the potential difference between the two lines. From these readings, ΔV_1 and ΔV_2 are determined; ΔV is the change in V when the current is interrupted, or the difference between the "on" and "off" values. These two give the first circuit constant,

$$B_{21} = \Delta V_2/\Delta V_1 \tag{10-1}$$

which is known as the ground voltage coupling. Note that this is a ratio between two voltages, and thus has no unit. It merely states how the potential of the foreign line responds to potential changes on the protected line. The same value should be obtained for this constant for any value of rectifier current, within a reasonable range; it is only convenient, not necessary, to use the value which will just protect the line.

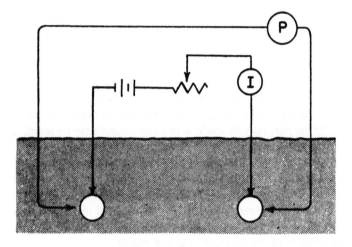

Figure 10-5. Zero-resistance ammeter. To determine the current which would flow through a "solid" or zero-resistance bond between two structures, the circuit illustrated is used. The current from the battery is adjusted until the potentiometer (or high-resistance voltmeter) reads zero. Then the current indicated by the ammeter I is the sought for value. There are instruments available which incorporate this complete circuit within themselves.

The next step is to open-circuit the rectifier, so as to remove its influence completely, and install a source of current, with interrupter, at the terminals (1). A storage battery, or, in many cases even a single dry cell, will usually serve this purpose. This makes it possible to transfer current from one line to another, as will later on be done by the bond, and to determine the effects of bond current. The interrupted current and the voltage across the terminals (1) between the two lines are measured; readings are also made of the *P/S* potential of the foreign line at terminals (2). These last two readings are called V_1 and V_2, respectively, but these are *not* the same values used before; they merely happen to be taken at the same points. From these three sets of readings, we get two more circuit constants.

$$R_{21} = \Delta V_2 / \Delta I_1 \qquad (10\text{-}2)$$

and

$$R_{11} = \Delta V_1 / \Delta I_1 \qquad (10\text{-}3)$$

The second of these, R_{11}, can be recognized as the coupling resistance between the two lines. R_{21}, since it is the ratio of a voltage to a current, also has the units of a resistance, but it does not correspond to any actually existing resistance; it is called the *transfer resistance*. I_1, when the interrupter is open, is equal to zero, so the value of ΔI_1 is just the value of current used.

From these three circuit constants, the correct value of the bond resistance can now be calculated:

$$R_B = \frac{R_{21}}{B_{21}} - R_{11} \qquad (10\text{-}4)$$

Since the value of the rectifier current does not appear in this expression, it would seem that the bond resistance is independent of the drainage current. This is indeed the case, but *only in the simplified case in which the two line potentials are equal*. When, as is almost always the case, these potentials differ, the bond resistance is given by

$$R_B = \frac{R_{21}(E_1 + \Delta V_1)}{B_{21}\Delta V_1} - R_{11} \qquad (10\text{-}5)$$

where E_1 is the natural potential between the two lines, and ΔV_1 is the potential change at terminals (1) caused by the interruption at the rectifier of the adopted rectifier current.

The fact that the bond resistance depends on the rectifier current now poses a problem. For, if the rectifier is adjusted so that it just protects the line, and the bond is then installed, it will be found that the line is no longer protected because the bond is taking part of the current. If, then, the current is increased to where the line is once more protected, it will be necessary to lower the bond resistance, and protection will again be lost. This leads to an endless pursuit of a properly balanced system, although actually the approximation becomes closer with each change and any three or four adjustments will usually suffice.

There is, however, an easier way, in which it is all done at once. First, the rectifier is adjusted to a value which is somewhere in the neighborhood of full protection; the exact value is not important, but it should not differ too much. A survey is then made to determine the point of maximum interference; this is the point (2) in Figure 10-4. Then the current source is connected between the two lines and a similar survey conducted along the protected line, locating point (4). The current to be used in this survey should be an approximation of the finally adopted bond current. As this is impossible to know in advance, a guess must be made. If the guess turns out to be wrong by too great a factor, it may be necessary to repeat the test. Point (1) is, as before, the tie between the two lines, and point (3) is the rectifier drain.

With the rectifier current being interrupted, the following network constants are determined: R_{13}, R_{23}, and R_{43}. With the bond current being interrupted, the following are found: R_{11}, R_{21}, R_{41}, and B_{21}. E_1 is also measured (preferably first, to avoid errors due to polarization by test currents). From these circuit constants, the following values can be computed:

$$I'_3 = I_3 \frac{R_{43}R_{21}}{R_{43}R_{21} - R_{23}R_{41}} \tag{10-6}$$

$$R_B = \frac{(R_{13}I'_3 + E_1)R_{21}}{R_{23}I'_3} \tag{10-7}$$

I_3 is the value of rectifier current found to be adequate without any bond current; I'_3 is the value which will be necessary in order to provide for the bond. R_B is the resistance of the bond which will eliminate the interference when the final adjustment to the current I'_3 has been made.

In all of these discussions, if the value of the bond resistance is made *less* than the calculated value, there will be some cathodic protection of the foreign line near the crossing; if it is made more, the interference will not be completely eliminated. The final test, after all the adjustment and installation is complete, is this: take the *P/S* potential of the foreign line at the point of maximum interference, with the rectifier operating and the bond in place; then simultaneously open-circuit the rectifier and the bond. The potential of the foreign line should not change if the adjustment is perfect.

Multiple Bonds

Equations similar to those given have been derived for the simultaneous adjustment of two or more bonds to different foreign lines and for the correction of rectifier output to compensate, but they are complicated and cumbersome in use. The following procedure is simpler and almost as accurate:

1. Adjust the rectifier so as to give adequate protection to the line, with no bonds installed. With this current being interrupted, locate points of maximum interference for all of the crossing lines (potential differences between lines should be determined before rectifier current is started).
2. Impress current across the first bond location, locate point of maximum bond effect, and make the measurements required for bond adjustment, just as in the single bond case above. This will give a value of I'_3 from which a certain amount of additional rectifier current is seen to be required for the first bond.
3. Repeat this step for each of the other bonds, keeping the rectifier at its original value. Each of these will give a value of added current required.
4. Add up all these extra requirements, and add this sum to the basic rectifier current. This will give the final value needed, and, when used as I'_3 in Equation 10-4, will give the resistance value for each bond.

When this has been completed, a complete survey should be made to see that the line is actually adequately protected. If not, the current must be increased, which may call for additional bond adjustments; these should be minor.

Auxiliary Drainage

In many cases, such as that illustrated in Figure 10-6, the point of exposure on the foreign line is not located at a point of crossing (the same is true of that in Figure 10-2). In such cases, often the best solution is the installation of one or more magnesium anodes at the point of exposure, thereby affording sufficient cathodic protection, or auxiliary

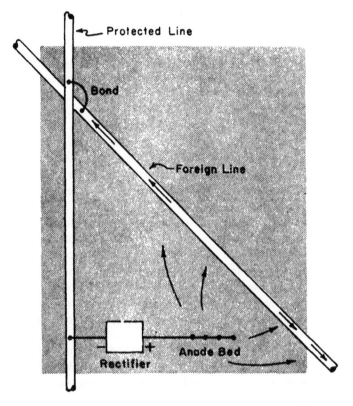

Figure 10-6. An obscure case. Much of the current collected on the foreign line flows toward the crossing, where it can be safely handled with a simple bond. Some of it, however, flows in the opposite direction and is discharged over a relatively large and remote area. Such a situation does not arise often and probably does little damage in any case because of the large discharge area. It can be avoided by proper anode bed placement and remedied by the use of auxiliary drainage anodes.

drainage, to the affected structure to avert the damage done by the offending system. In effect, this offers a path to ground for the collected current which can be taken without damaging the structure—the damage, instead, is inflicted on the magnesium. This same technique can be used in simple crossing cases, instead of the bond described above; the collected current, instead of following the bond back to the protected line, flows to earth by way of the magnesium anodes.

Parallel Lines

When cathodic protection is applied to a pipe line which occupies the same right-of-way with another unprotected line, there exists an opportunity for the foreign line to collect current in some areas and discharge it through the earth to the protected line in others. The determination of the areas in which such discharge takes place can most easily be made by a modification of the surface potential survey previously discussed. In the present case, two electrodes are used, one being placed over each line, for the first reading, and one over the foreign line and the other offset a distance equal to the space between the two for a second reading, as shown in Figure 10-7.

Readings are to be taken in both positions with the unit on and off, and differences used. By this means it can be determined whether there is current flow from the foreign line to the protected line, as distinguished from current flow past the foreign line to the protected line. When the areas of undesired flow are located, bonds should be installed between the two lines at the points where the conditions are the worst.

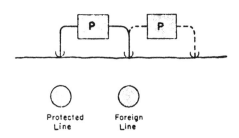

Figure 10-7. Current transfer between parallel lines. If the delta (difference between on and off readings) in the position shown in solid lines is appreciably greater than that in the dotted position, then there is current transfer from the foreign line to the protected line. When the point of worst exposure is located, a bond should be installed. A repeat survey must then be made to determine the length of section which the bond will protect, and other bonds installed if required.

Further tests will then be needed to see if the bonds are adequate; it is not possible to determine in advance whether a given bond will take all the current discharge over a section, or whether several will be required.

Radial-Flow Interference

As indicated in Figure 10-2, it is possible for a structure to suffer damage by interference currents when it does not lie close to the protected line. The type of exposure illustrated is known as *radial-flow interference,* which means merely that the structure lies in an area where there is current flowing through the earth; since it lies at least partly in the direction of flow, it picks up some current on the side next to the anode and discharges it on the far side. Such trouble is not likely to be encountered anywhere except very close to an anode bed and it is most easily remedied by the installation of auxiliary drainage, as described earlier. The installation of anode beds remote from structures will avoid problems of this nature, but this is by no means always possible.

Foreign Lines with Insulating Joints

Figure 10-8 shows another difficulty which may be encountered. In this case the foreign line has nonconducting couplings, such as Dressers; the current collected cannot be conveyed to the point of bonding without having to bypass the coupling. At each bypass point, some current flows off the line, through the soil, and back to the next joint, with attack on the side of the coupling away from the protected line.

Such a line, remote from the anode bed, would not pick up enough current to cause significant damage; only when the line lies in a heavy current density region is there much likelihood of damage. The damage may be forestalled either by bonding the poorly conducting joints (and the installation of a crossing bond) or by the installation of auxiliary drainage on each joint of the line through the exposure area. The two methods may be combined, by the installation of a bond and a magnesium anode on *alternate* couplings, so that the line is separated into two-joint sections, each one of which has a single auxiliary drainage anode.

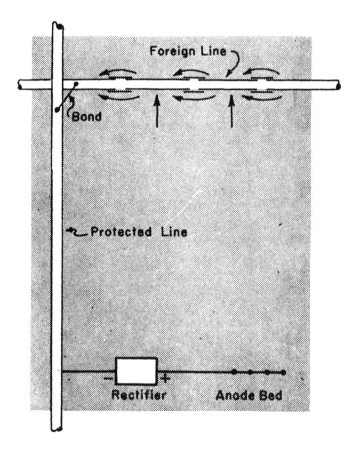

Figure 10-8. Foreign Line with Dresser Couplings. The installation of a bond at the point of crossing will avert the damage there, but there will still be damage done at the mechanical joints, by current bypassing them through the earth. This can be remedied by bonding the joints or by the use of auxiliary magnesuim anode drainage.

Summary

As has been shown already, interference problems are always a possibility whenever a cathodic protection system will be installed. This system will affect:

1. *Foreign lines.* These include protected lines and those with their own protective systems. These interference problems are usually solved by resistance bonds.
2. *Insulating flanges.* To prevent the current necessary for protecting a line from being drained off on such piping systems as tankage lines, well casings, and unused pipe lines, it is sometimes necessary to install insulating flanges or Dresser couplings. Corrosion will occur on the buried piping near these flanges unless bonds are placed around the flanges and adjusted to balance corrosion forces.
3. *Railroad tracks and road casings.* Any metallic system in the vicinity of the protected pipe line may cause problems. Each of these will be shown by accurate measurement and must be solved as it occurs.

11

Operation and Maintenance

After a cathodic protection system has been installed, there are certain methods which we can use to evaluate its performance. Its major justification will be that corrosion has been prevented, but no signs are outwardly noticeable. The purpose of this chapter is to describe what procedures to follow to assure that the systems are in operation.

Importance of Adequate Supervision

The total or partial failure of a cathodic protection system is not accompanied by any outwardly visible sign. If a pump fails, there is a pressure drop or a pressure rise somewhere, and the condition will call itself to the attention of those responsible. Much the same thing is true of most electrical or mechanical systems installed to perform various functions. But if there is a failure on a cathodic protection system, the pipe-to-soil potential falls (and this is not visible, except to the specialized "eyes" of instruments), and the relatively slow processes of corrosion resume their damaging activity. Then, unless someone is on the alert, there is no indication whatever of the failure until a leak occurs. By this time, a great deal of damage has been done to the supposedly protected system, in addition to that done at the one point of maximum corrosion rate, where the leak has occurred.

Besides the fall in pipe-to-soil potential, there are some other indications of failure which may come to the attention of the supervisory forces. If the system employs a rectifier, there will usually (but not always) be a change in the readings of the ammeter and voltmeter on the unit. There will also be, if the failure is in the unit itself, a drop in the total power consumption—which will show up on the monthly watt-hour meter readings and the power bill. If the system is based on magnesium anodes, there may be a change in the current output of the anode groups; but this is no more likely to be observed than is the fall in pipe-to-soil potential. The different things which can happen to produce total or partial failure of the protection system can best be examined by a separate consideration of the two principal types of systems, rectifier-ground bed systems, and magnesium anode systems.

Failures in Rectifier-Ground Bed Systems

It must be remembered that a failure can occur anywhere in the complete system, which includes the protected structure as well as the protective devices. When the protection is by means of a rectifier draining current to a ground bed (see Figures 11-1, 11-2, 11-3, and 11-4), the following parts of the total system are subject to failure:

1. The power supply; failure may be permanent or temporary.
2. The rectifier—transformer, stacks, circuit breaker, or auxiliaries.
3. Cable connection to the ground bed, or between elements of the ground bed; thus the loss of protection may be complete or partial from this cause.
4. The ground bed itself; one or more anodes may become disconnected; there may be an increase in total resistance due to a change in soil moisture conditions; or there may be total or partial consumption of the anodes.
5. Cable connection to the pipe itself.
6. Accidental electrical connection to the protected structure of a mass of metal whose protection is not contemplated, thus overloading the system; this may be an accidental contact with an old or new line, a short-circuited insulating joint, or a contact with a road crossing casing.
7. Deterioration of pipe coating with time, or by damage, so as to increase the demand of the system above the capacity of the unit installed.

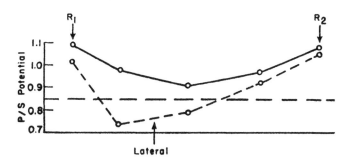

Figure 11-1. Pipe-to-soil potential before and after failure. The solid line shows a typical pipe-to-soil potential plot between two rectifiers with the line under adequate protection. The dotted line shows a survey over the same section after an insulated joint has become short-circuited, throwing onto the line an unprotected lateral. The maximum effect is seen to be at the points adjacent to the lateral.

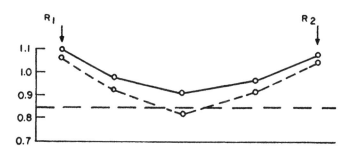

Figure 11-2. Pipe-to-soil potential after coating deterioration. The solid line shows the same section illustrated in Figure 11-1; the dotted line shows the potentials found along the section after a uniform deterioration of the pipe coating. The effect is seen to be distributed.

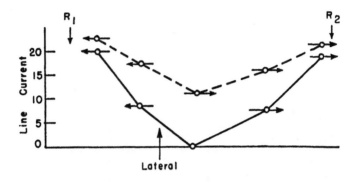

Figure 11-3. Line currents as affected by failure. The dotted line shows the values of line currents in the same section illustrated in the two preceding figures. The dotted line shows the line currents after short-circuiting of the insulated joint to the unprotected lateral. The effect is more pronounced than that obtained by pipe-to-soil potentials; further measurement of line currents in the vicinity of the lateral would be even more definite.

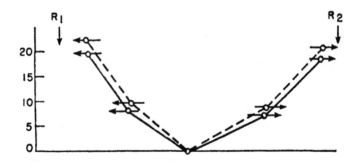

Figure 11-4. Line currents as affected by coating deterioration. As indicated, using the same conventions as in the other three figures, the effect on line current distribution of a general coating deterioration is very slight; note that the potential is affected much more severely.

Some, but not all, of the listed failures will affect the readings of the meters on the rectifier unit. It should be noted, however, that *all* failures—total or partial—will show up as a loss in pipe-to-soil potential in at least part of the system.

Failures in Magnesium Anode Systems

The cathodic protection system as a whole is a more complex network when multiple drainage with magnesium anodes is used; this may mean that some types of failure will affect only small portions of the system. It also means that more points must be checked to be sure that the entire network is functioning. Systems employing magnesium anodes are subject to failures of the following kinds:

1. Consumption of one or more anodes by normal use and by local action. The effect will be a falling off of both current and potential.
2. Mechanical damage to connecting wires.
3. Loss of current output due to abnormally dry soil; this effect may be produced by dryness at either the anode or the pipe, or both.
4,5. Same as numbers 6 and 7 above: i.e., effects connected with the pipe itself.

As with the rectifier system, some of the failures will show up as failure of current output at the anode; there is no indicating meter on which this can be read, but the current in the lead can be measured if a shunt has been installed, or the lead may be cut for the insertion of an ammeter. Again, any of the failures will produce a loss in pipe-to-soil potential.

Minimum Inspection Schedule for Rectifier System

It is suggested that the voltmeter, ammeter, and watt-hour meter of each rectifier be read daily, weekly, or as a minimum, monthly. The more frequent readings are advised only where accessibility is easy, such as with units installed at manned stations or where daily visits are made for other purposes. Frequent readings are useful for two reasons: first, trouble is found and remedied in less time, so that less protection is lost; and second, the shorter the interval in which the trouble has been known to develop, the easier it is to guess at probable causes.

These readings should be plotted on a graph, which makes trends easier to observe. Gradual changes may be either seasonal variations in soil moisture or coating deterioration, or even rectifier stack deterioration. Sudden changes are the ones which most need investigation and correction.

In addition to the routine readings described, a more thorough rectifier inspection should be made at greater intervals; different companies use quarterly, semiannual, and annual periods for this inspection. The following items should be included:

1. Before opening the case, look for damage or tampering; is the watt-hour meter operating? Feel the case for warmth and listen for the low hum of an operating rectifier.
2. Open the door carefully. Look out for snakes, spiders, insects; read the meters, note the settings, and record, along with the watt-hour meter reading. Look for signs of damage or trouble.
3. Turn the power off by the outside switch, so the whole unit is dead. Immediately, feel all of the stacks, they should be warm, depending on the load. More important, they should all be at about the same temperature. A cold stack is not working; if two are cold, probably only one of them is defective.
4. Without pause, feel all of the connections that can be reached. None of these should be hot, or even very warm. Any found hot or warm should be tightened, or, if found tight, should be dismantled, cleaned and/or filed smooth, and retightened.
5. A more leisurely inspection should now be made. Look for any burned places from possible arcs or lightning damage. Check the unit for supply of fuses; check operating fuses for proper size.
6. Inspect screens for stoppage by dust, insects, bird nests, or other debris; clean if needed.
7. With the unit still off, check the watt-hour meter; the disc should be absolutely motionless.
8. Check the panel meters by comparison with (portable) meters of higher precision.
9. Inspect the case for possible repainting. Look over the whole installation: pole, guys, cable runs, warning signs, and any other accessory equipment.

10. Reenergize the unit and check efficiency. This is the ratio of DC output to AC input, expressed as a percentage. The DC output is the product of the volts and amperes, as shown by the panel meters (corrected as in step 8).

The AC input is determined from the watt-hour meter, by counting the revolutions in a given time, or the time of one revolution if the load is very light. Output in watts is then given by

$$P = \frac{3600NK}{T} \qquad (11\text{-}1)$$

where:

P = AC input, watts.
N = number of revolutions of the disc.
K = meter constant (found on the nameplate of the meter).
T = length of the counting interval, seconds.
The efficiency is then computed from

$$\text{Eff} = \frac{IE}{P} \, 100 \qquad (11\text{-}2)$$

This value should be plotted as a function of time. Any unit will eventually show a decline in efficiency as a result of stack aging. As a general rule, when the efficiency has dropped as much as 25% from the original value (at a comparable load), the stacks should be replaced.

Rectifier efficiency can also be averaged over a month, or other period between meter readings, by the following expression

$$\text{Eff} = \frac{IET}{1000W} \qquad (11\text{-}3)$$

where:

I and E = meter values averaged over the period.
T = length of time, hours.
W = total power consumption for the period, kilowatt-hours.
Even when a rectifier is not visited at all, there is some value in making this check by using the power company's billing values. A pronounced change in efficiency or a totally inoperative unit will be found.

Minimum Inspection Schedule for Anodes

It is recognized that anode installations cannot be inspected as thoroughly or as often as can rectified systems, because of the larger number of components. Then, too, very often the anode type of installation is on lines which are of lesser economic importance. The general technique is much the same; there are no power meters to read, however—nothing but anode currents and pipe-to-soil potentials. Shunts installed in the anode leads make the reading of current a comparatively simple matter, though not as simple as the reading of a permanently installed instrument, as found on the rectifier. It is suggested that a sampling technique be used, and pipe-to-soil potentials at about one-third of the midpoints be checked monthly; the other midpoints, and the current outputs, may then be checked semiannually or annually. Total current output in ampere-hours should be computed so that the estimated life of the anode may be anticipated and replacements scheduled as needed.

Monitor System

A technique which may be applied to either the rectifier or galvanic anode system is that of using the potential monitor. This is an instrument which is permanently installed at a number of critical points, so selected that they are indicative of the behavior of the system as a whole. These should be somewhere near the midpoints of sections between rectifiers, but must be at easily accessible locations. Since they can be read by unskilled personnel, it is possible to use field men who are in the general area; readings are mailed or called in to the corrosion department.

The instrument consists of a medium-resistance voltmeter, usually 1000 ohms per volt, installed in a suitable protective housing and connected so as to indicate the potential between the pipe line and a buried zinc anode. Although this voltage is not the same value as that with reference to the usual copper sulfate electrode, the two can be correlated. One form of the monitor uses a zero-center voltmeter, reading 500 millivolts in each direction. Another form, commercially available as a complete unit, has essentially the same meter but is marked so as to indicate copper sulfate potential directly; provision is made for adjustment.

Figure 11-5. Potential monitor. (Courtesy of M. C. Miller Co., Inc.)

These values should be read at small intervals, and should be recorded and plotted. There will usually be a slow cyclic variation with the seasons, an irregular variation with soil moisture conditions, and perhaps an overall slow decline connected with coating deterioration. A sudden drop is definite indication of some kind of failure and calls for an investigation. Figure 11-5 shows a typical test station potential monitor, which is usually installed at the midpoint of a pipeline and which has a zinc anode calibrated to give readings similar to those secured from a copper sulfate electrode.

Troubleshooting

Rectifier-Ground Bed

When the monitor, or a pipe-to-soil potential survey, indicates inadequacy of protection, the first place to look is at the protective unit. The current output of the rectifier should be checked; if it is normal, the trouble is on the line itself; if it is high and accompanied by low voltage, the trouble is certainly on the line and is caused either by increased current demand or by a short-circuit to parasitic metal. If the current output is low, with voltage normal or high, the trouble is in the ground bed or connecting cables. A pipe-to-soil potential over the ground bed will show a peak over every anode which is working; a disconnected anode will not show at all. This test is particularly useful if comparison can be made to a similar test made at the time of the original installation. When inactive anodes are found, only digging will uncover the cause.

If the rectifier and its anode bed appear to be performing satisfactorily, the source of the low potentials must be sought on the line itself. Any area in which work has been done recently should be investigated; for example, if a new lateral has been connected, the insulation should certainly be checked. If investigation of such suspected sources discloses nothing, then a more detailed search must be made. First, the pipe-to-soil potentials should be studied, to see if the failure seems to be localized. A more thorough, but slower, approach is that of making a detailed line current survey; find out where the drained current is coming onto the line. This will be much more easily interpreted if a similar survey, made when the line was in satisfactory condition, is on record; a comparison will often very quickly locate the offending parasite. On the other hand, if the current collected in each section is greater than on the earlier survey, with no pronounced differences, then the trouble is simply that of increased overall current requirement, probably due to coating deterioration.

Magnesium Anode System

The basic technique is the same as that outlined before; more measurements are required, because of the multiplicity of drain points (see Figure 11-6). The current output of stations nearest the point of low potential should be checked; if these are satisfactory, a similar check should be extended in both directions until it is clear that the trouble must be on the line. When a given anode group shows a marked drop in current output, the cause may be drying out, shrinkage of backfill, or severed or broken lead wires. If the current is zero, the pipe-to-soil potential of the lead wire will show whether it is still connected to the pipe or the anode, and thus indicate the direction to the failure. If the current is low, there may be a loss of one or more anodes by a severed wire; a pipe-to-soil potential survey over the anodes will show which are active, just as in the case of rectifier anodes.

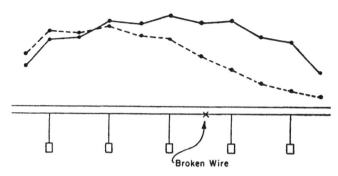

Figure 11-6. Locating idle anodes by surface potentials. The solid line shows the potentials found along a line of anodes when all are delivering current; the dotted line exhibits the change when there is a break in the anode lead at the point indicated. Single disconnected anodes may also be located by this method. A driven ground rod, a pipe lead, or even a rectifier terminal may be used for the reference ground; all readings should be referred to the same reference.

Trouble indicated as being on the line, rather than at the anode stations, is tracked down in the same manner as that used for lines protected by rectifiers; first investigating locations where something has been done which might be responsible, next checking potentials, and finally making a line current survey.

Summary

Once the cathodic protection system is installed on a pipe line, it must be kept in operable condition at all times. Certain routines are necessary to be sure that the rectifiers are running and that the anode bed leads have not been severed. A minimum inspection system has been shown to maintain adequate performance. This is even more important when a sacrificial anode system is used, since there is no rectifier to show inadequate performance. Since pipe-to-soil potential measurements must be run at regular intervals along a pipe line, it is some advantage to have a potential monitor station, usually a voltmeter with a zinc anode, installed at certain critical points on a pipe line. However, a yearly complete cathodic protection measurement determining line currents, potentials, and coating conductivities should be run.

12

Coating Inspection and Testing

Construction Inspection

It often becomes the duty of the corrosion engineer to inspect coating in connection with a pipe line construction project or to supervise such inspection. To do this, he must be familiar with the general types of coatings we have already discussed in Chapter 8. These coatings are applied to the pipe line in two general methods: (1) in the shop, and (2) over the ditch.

In the Shop (in 30-foot sections)

Ideally, the pipe is sandblasted to a white-metal surface. Next, it is primed with an air-curing coating that serves as a base for the top coat, which is then applied to a thickness between 2 and 3 mils (0.001 inch). If it is a bituminous coating (either coal tar or asphalt), it may be added in a molten state. If it is a thin-film epoxy, it is cured by the addition of a peroxide. Then it may be protected for transportation by a paper-wrapping machine. The major problem for shop-applied coatings is joint protection after welding on the job site.

Over the Ditch

The pipe has been laid out next to the pipe line excavation. It is then welded, cleaned by an automatic wire-brush machine, and immediately primed with an air-drying coating. Then the molten bituminous coating is applied and followed by a wrapping of glass fiber and Kraft paper. The advantage is that it is now a continuous system. If the pipe line is to be taped, the procedure is not complicated by the necessity of the heating kettle and may proceed faster.

The inspection for the shop-applied coating must be done twice—once at the shop and again on the job site just before burying the pipe. The over-the-ditch coating job may be inspected only once—just before burying the pipe.

The instrument used for finding the coating defects (''holidays'') is the electrical holiday detector. This consists of an electrical energy source such as a battery or a high-voltage coil, an exploring electrode, and an alarm to signal current flow through the apparatus. When the apparatus is used for pipe line coating testing, the electrode is composed of a full circle spring to surround the pipe. When the apparatus is used for large areas such as tanks, it uses a wand with a brush electrode. The voltage may vary as much as 30,000 volts for a ''spark'' detector for thick bituminous coatings and a maximum of 75 volts for a ''nondestructive'' detector for thin coatings. Figure 12-1 shows the instruments in use.

The ring spark detector is grounded and passed along the pipe line. Any imperfections in the coating will cause a spark to form, and this will be amplified by a bell signal. This is the usual method of operation. It should be remarked that this full-circle detector cannot always be used and that a semicircle detector is also needed.

There are at present two fairly distinct schools of thought on the subject of coatings for major transmission lines. One of these holds to the belief that the coating should be made as nearly perfect as possible; from this point of view, every foot of the line should be carefully checked with the detector, and every holiday found should be carefully repaired. The adherents of this view often follow the finished construction with a complete foot-by-foot Pearson survey and repair all defects thus found. The thinking behind this view is that the line, when completed, will be so nearly perfect that a very small amount of cathodic protection will be required. This is valued, not so much for its low cost, but because a nearly perfect line under cathodic protection is very sensitive to any disturbance; even a few square inches of damaged coating, miles from the nearest unit, will produce a detectable change in the po-

(a) Full Circle.

(b) Semicircle and Full Circle Equipment

Figure 12-1. Use of the holiday detector.

tential. Hence such a line may, by careful supervision, be maintained in excellent condition for a long period of time.

The opposing viewpoint is that the cathodic protection of a line with reasonably good coating—not the superb coating just described—costs far less than the extra cost of the superior coating. Adherents of this school seldom make 100% holiday detector inspection, do not patch small holidays, and use Pearson surveys only on sample sections of the line. They also usually specify less-expensive combinations, omit some reinforcing materials, or rock shields, and choose various other economies. The current demands of the line may then run from four to ten times as high as those with more rigid requirements, but the claim is made that this extra cathodic protection is less expensive than the more elaborate coating job.

Before attempting to pass judgment on this difference of opinion (which, after all, lies outside the scope of the field engineer), it should be noted that the possible cost of a leak varies widely on different lines; and, further, that the financial and tax positions of various companies are quite different, and this may strongly influence the amounts they are willing to spend on capital investment and on maintenance charges.

Evaluation of Coating in Place

The coating on a buried line may be evaluated in any one of three ways: by a current requirement test, as described in Chapter 4; by the measurement of coating conductance, or, more properly, leakage conductance; or by locating breaks and holidays with a Pearson survey. Visual spot inspection is of little value, except in the case of very poor coatings or very good, and then the results are inconclusive.

Coating or Leakage Conductance

The leakage conductance of a section of pipe is expressed in micromhos per square foot. It is also common practice to express the conductance of a line in terms of length rather than area, in which case the unit is micromhos per foot. Many conductance calculations are made using as a unit of length the megafoot, or one million feet. In this system the conductance is expressed in mhos; mhos per megafoot is actually the same as micromhos per foot. If a change in potential with respect to remote earth of one volt produces a change in the leakage current of a line section of one ampere, then the total conductance of that section is one mho. Conductance per unit length or per unit area can then be obtained by dividing total conductance by length or area.

Values obtained in practice range from as low as 1–10 micromhos per square foot (excellent coating in high resistivity soil), 10–50 (good coating in high-resistivity soil, or excellent coating in very low-resistivity soil), up to as high as 2000 or more (very poor coating); even bare line has a measurable "coating" conductance, which, like that for coated pipe, is strongly influenced by the soil resistivity.

The National Association of Corrosion Engineers Technical Practices Committee T-10D-1 has revised the NACE Report 2D157 *Methods for Measuring Leakage Conductance of Coating on Buried or Submerged Pipelines* into an NACE Standard RP and gives an outline of methods of measuring this quantity. The description is general and requires considerable elaboration for actual field use. Three different methods, applicable in three different cases, are given here:

1. *General method.* This is based on measuring directly the basic quantities involved; i.e., the total current flow to a given section, the average potential shift in that section due to the current, and the length of the section. The arrangement used is illustrated in

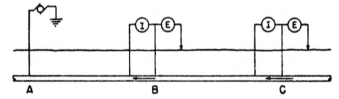

Figure 12-2. Coating conductance measurement. An interrupted current is drained at A. At B and C, measurements are made of the line current with drainage on and off; the difference is ΔI. At the same points, measurements are made of pipe potential with respect to a remote electrode with drainage on and off; these difference are ΔE. The total current picked up in the section is then $\Delta I_B - \Delta I_C$; the average potential shift is $\frac{1}{2}(\Delta E_B + \Delta E_C)$. Conductance of the section is then I/E, or

$$K = \frac{2(\Delta I_B - \Delta I_C)}{\Delta E_B + \Delta E_C} \quad (K \text{ in mhos})$$

This value divided by the area of the section gives the conductance per square foot.

Figure 12-2. Two cautions are to be observed in using this method: the two values of ΔI must differ significantly (e.g., if one value is 6.25 amperes and the other is 6.20, the difference of .05 is not known with sufficient precision; a ratio of at least 2:1 is recommended). The other factor to observe is that the two values of ΔE must not differ by too much, otherwise the arithmetic mean is not a sufficiently close approximation to the true average potential shift. The ratio of the two values of ΔE should not be greater than 1.6:1, in general. If these limits are not met, the interval must be shortened or lengthened, as needed.

2. *Long line method.* When the line section is sufficiently far from the end that the behavior is that of a *long line* or a *very long line* (see Chapter 5), then a simpler procedure is applicable. An interrupted current is drained, and readings of *P/S* potential are taken at two points—one far enough from the drain point to avoid the anode proximity effects, the other as much farther as is desired, so long as the reading is large enough to be usable. (Note that it is usually necessary to take more than two sets of such readings to determine whether the line is a *long line*, etc.) No current readings are required, except to be sure that the current is the same for both of the potential readings. The attenuation constant can then be obtained from

$$\alpha L = \log_e (\Delta E_a / \Delta E_b) \tag{12-1}$$

where

    α = attenuation constant
    L = distance between the points a and b, at which ΔE is measured
  $\log_e$ = natural logarithm—to be taken of the potential ratio
From this value, we can then obtain k from

$$\alpha = \sqrt{rk} \tag{12-2}$$

where

    r = longitudinal resistance of the pipe
    k = conductance per unit length; the megafoot is a convenient unit to use

If calibrated current measuring spans are available, the same method can be applied, using line currents instead of potentials; the first expression is then replaced by

$$\alpha L = \log_e (\Delta I_a / \Delta I_b) \tag{12-3}$$

and the rest of the calculation is the same as before. Line currents are often subject to much less outside influence than potentials, so this method is often more accurate and consistent.

3. *Terminated line method.* If the ΔE at a termination of the line (insulated joint or open, disconnected line) is large enough to be useful, then it and one other value of ΔE may be used to compute the attenuation constant, and thus the conductance. The other value must be far enough from the drain point to avoid the anode proximity effect, and must be different enough from that at the end to give a usable ratio between them. The expression is

$$\cosh \alpha L = \Delta E_a / \Delta E_L \tag{12-4}$$

where:

   cosh = hyperbolic cosine, tables of which are available
   α and L = two points referred to where the potentials are measured
   L = distance between them

The rest of the calculation is the same as the previous one.

4. *Other methods.* There are at least a dozen other methods of determining coating conductance; one who has a thorough knowledge of the attenuation equations and the mathematics involved can often save a great amount of field work, installation of test leads, etc., by choosing the method best adapted to the situation at hand.

Pearson Surveys

The apparatus devised by Dr. John M. Pearson permits the location of breaks and holidays in the coating of a buried line. The principle involved is that of impressing an alternating voltage between the pipe and the earth, and then detecting the high potential drop in the neighborhood of a bare spot. A signal generator (usually a vibrator) is connected to the

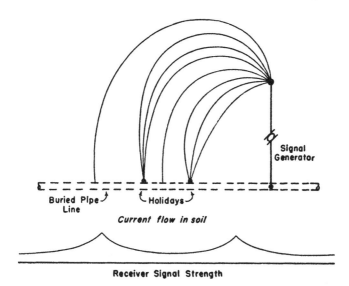

Buried Pipe
Line

Holidays

Current flow in soil

Signal
Generator

Receiver Signal Strength

Figure 12-3. Pearson holiday detector. The current flowing from the signal generator ground to the pipe line concentrates at the holidays; this produces a concentration of current in the ground, which results in peaks in the receiver signal, as indicated on the graph.

line, the other terminal being connected to a ground rod a few hundred feet away. Then a team of two men walk the line about 20-feet apart. Each man wears a pair of contact plates on his shoes; the potential difference between two points 20-feet apart is thus picked up. This is fed to an amplifier carried by the front man, and the amplified signal can be heard in earphones and is indicated on a meter. The rear man handles the connecting cord, keeps it clear of obstructions, and maintains reasonably constant distance. Figure 12-3 shows a typical Pearson reading, while Figure 12-4 shows a pictorial representation.

As a holiday is approached, there is a rise in signal intensity which reaches a maximum when the front man is directly over it. Another maximum is heard when the rear man passes the same point. On lines with many holidays, confusion may be avoided by having one man walk the line while the other walks parallel to him at a distance of 20 feet; in this way each holiday has only one signal.

Figure 12-4. Pearson holiday detector operation. The concentration of current strength in the soil, in the vicinity of a holiday, is picked up by the shoe contacts worn by the two men and amplified by the apparatus carried by the operator.

The operation is quite similar to that of conducting a surface potential survey on a cathodically protected line. Alternating current is used instead of direct, and a different system of detection is employed. The results obtained by the surface potential survey are much more quantitative; the Pearson technique, however, is a great deal faster, and there is little or no danger of skipping a holiday, since it is an essentially continuous operation.

Accelerated Coating Tests

Numerous tests have been devised for the comparative evaluation of different coating materials and of different coating techniques. These consist of applying the materials to be tested to short lengths of pipe or other metal specimens, and then subjecting them to highly corrosive exposures, often with externally applied voltages; the specimen is sometimes made the anode and sometimes the cathode. All of these methods are subject to the criticism that they do not accurately reproduce field conditions and that they are subject to the errors of small samples.

It is difficult or impossible, however, to compare coatings or application methods on the basis of actual pipe line conditions. Long periods of time are required, and it is almost impossible to find two sections of line which can be considered identical. Also, the periodic inspection and testing needed for data is not only expensive, but it disturbs the sites so that even field tests do not duplicate ordinary field conditions.

Accelerated tests are useful in eliminating very poor coatings or techniques, and in pointing out those which show promise of excellence in actual use; experience with operating lines must then give the final answers, and it cannot be expected that they will be had in a short time, or that any but statistical averages of a large number of lines can yield valid results. By the time such tests have progressed far enough to give dependable data, new developments may have produced new coatings which ought also to be tested in this field.

Summary

It has just been shown that the best way to assure a good coating system is by adequate inspection at the time the pipe line is constructed. The holiday detector is an efficient instrument to find cracks in the coating, but it does not determine the bonding of the coating to the pipe surface. This must be done by careful visual inspection. After the pipe is buried, several methods of determining coating conductivity are available, and it has been shown that this characteristic can be used in the design of a cathodic protection system. The instrument often used in looking for coating faults is the Pearson holiday detector, which requires two people to make the measurements. After this inspection, the pipe may be excavated and the coating repaired.

Appendix A
Fundamentals of Corrosion

Corrosion, according to the National Association of Corrosion Engineers, is the deterioration of a material caused by its environment. Corrosion control is thus the prevention of this deterioration by three general ways: (1) change the environment, (2) change the material, or (3) place a barrier between the material and its environment.

All methods of corrosion control are variations of these general procedures, and many combine more than one of them. The material does not have to be metal but *is* in most cases; the metal does not have to be steel, but, because of the strength and cheapness of this material, it usually *is*. Again, the environment is, in most cases, the atmosphere, water, or the earth. There are, however, enough exceptions to make corrosion control a bit complex.

By far, the classic example to study is iron in water. Actually, a piece of pure iron in distilled water in an oxygen-free atmosphere will not corrode. It is only the fact that we live in an atmosphere containing oxygen that even a beaker of distilled water in a laboratory will absorb oxygen from the air, and the iron will slowly go into solution with the chemical reaction

$$Fe + \tfrac{1}{2}O_2 + H_2O \rightarrow Fe(OH)_2 \text{ (ferrous hydroxide)}$$

It can be seen that both oxygen and water are necessary to attack the iron. Sometimes there are inorganic salts dissolved in the water, and they may increase the conductivity of the water and speed up the reaction even more.

The iron has undergone the following change:

$$Fe \rightarrow Fe^{++}$$

metallic   ferrous
iron   iron ion
solid   (in solution)

Consequently, a portion of the solid metallic iron has deteriorated into the liquid ferrous ion.

This is the fundamental corrosion reaction of iron and is similar to the section at the anode of an electrolytic cell in which iron is the anode$^{(-)}$ and a metal such as copper is the cathode$^{(+)}$. A classical cell has the structure:

1. An electrode which is destroyed (anode).
2. An electrode which is built up (cathode).
3. An electrolyte which will transfer ions.
4. A conductor between the electrodes to allow current to flow.

Figure A-1 shows this particular cell compared to the generalized concept of an electrolytic cell. This type of cell is called the *dissimilar electrode cell.* If a voltmeter is placed between the two electrodes, it will show the potential difference between the electrodes. Actually, because of differences in surface irregularities, the electrodes may be the same metal and a current will flow, destroying the anode and causing corrosion.

Two other types of electrolytic cells with the same metals for electrodes will also cause current to flow. These are:

1. Concentration cells
   a. Salt concentration
   b. Difference in oxygen content
2. Differential temperature cells

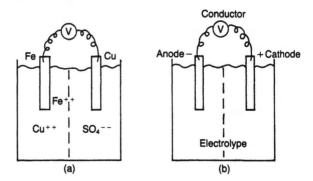

Figure A-1. Dissimilar electrode cells.

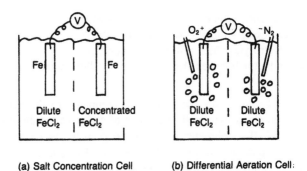

(a) Salt Concentration Cell   (b) Differential Aeration Cell:

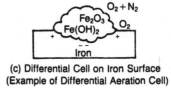

(c) Differential Cell on Iron Surface
(Example of Differential Aeration Cell)

Figure A-2. Concentration cells.

Figure A-2a shows a typical concentration cell in which the electrolyte is concentrated ferrous chloride around one electrode (cathode) and dilute ferrous chloride around the other. Current will flow from the concentrated to the dilute solution where the anode will be deteriorated to increase the concentration of ferrous ions.

Figure A-2b shows a differential aeration cell again containing two iron electrodes. The area around the cathode is subjected to oxidation by the admixture of oxygen, while the region around the anode is deaerated by bubbling in nitrogen.

Both of the previous examples are laboratory demonstrations of what can happen in a cell, but Figure A-2c is an example of a possible cell in a small crevice on the surface of iron. The rust product [$Fe_2O_3 + Fe(OH)_2$] serves as an electrolyte, while the cathodic area will be at the air surface and the anode will be in the anaerobic region beneath the rust. The iron will serve as a conductor. This accounts for the pockmarked surfaces seen on iron or steel in corrosive soil or water. Figure A-3 shows an example of a differential temperature cell where the only current generation is caused by the difference in temperatures around each electrode.

Thus, the electrolytic cell theory can be seen to offer a suggestion as to the corrosion of a single metal in a corrosive environment.

One caution should be mentioned. The data on electrolytic cells is based on the initial closed-circuit potential. As flow continues in a cell, a process called ''polarization'' occurs in which the potential difference between the cells begins to fall as the current increases. This characteristic will be of use when cathodic protection is used.

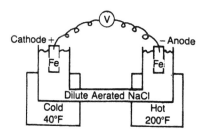

Figure A-3. Differential temperature cell.

Appendix B

Cathodic Protection of Steel in Soil

Cathodic protection of steel in soil is based on two general principles:

1. Steel corrodes because portions of the material in the soil are anodic and others are cathodic.
2. Corrosion will not occur if all portions of the steel are cathodic.

Therefore, the goal of cathodic protection is to make a cathode of the steel. This is done by impressing a direct electric current on the pipe and providing an anode which will corrode instead. This will not only reduce corrosion, it will stop it.

Cathodic protection is defined as the use of direct electric current to stop corrosion. This is done by examining the following principles of an electrolytic cell:

1. Only the anode corrodes.
2. The cathode does not corrode.
3. There must be an electrolyte for the currents.
4. There must be a connection between the two electrodes.

From a study of the various kinds of corrosion cells, we know that as soon as the cell starts to conduct current, a process called polarization takes place in which the potential difference of each cell is reduced until a quasiequilibrium condition exists. This cell equilibrium may exist in various states:

1. Anodic control (Figure B-1).
2. Cathodic control (Figure B-2).
3. Mixed control (Figure B-3).

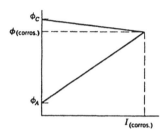

Figure B-1. Anodic control.

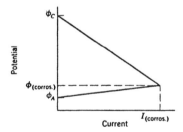

Figure B-2. Cathodic control.

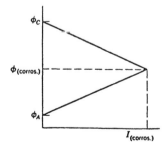

Figure B-3. Mixed control.

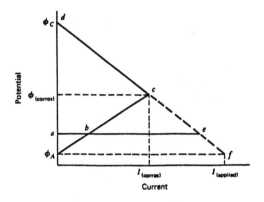

Figure B-4. Polarization diagram illustrating principle of cathodic protection.

By adding extra current from an external source, all cells may be placed under cathodic control, as shown in Figure B-4. This extra current may be identified as similar to the ΔI, which is necessary for cathodic protection. We have thus made the entire structure a cathode. To do this, we must have an external anode. If it is higher on the electromotive series than steel and the electrolyte can conduct the current, this may serve as a cathodic protection cell. If steel is to be used as the external anode, then a source of direct current must be found. This is usually done by using a rectifier (if alternating current is available).

The result will be that the protected structure is now polarized so that all the surface is cathodic and will not corrode.

Appendix C
Corrosion of Steel in Soil

The electrolytic theory shows an explanation of the corrosion of steel in water. Laboratory tests on steel in aerated water (Table C-1) show a rise in corrosion rate with increasing oxygen, up to a maximum at about 13 ml of oxygen per liter of water. Afterward, the excess oxygen is supposed to passivate the surface, and at 20 ml oxygen per liter, the corrosion rate is down to 2 mils per year, compared to 11 mpy at the maximum.

Like any chemical reaction, the corrosion rate of steel in aerated water doubles for every 55°F rise in temperature of the atmosphere in which it is confined. If the solution is allowed to boil in an open vessel, the oxygen boils off and the reaction rate falls.

The effect of pH on corrosion rate is constant (about 10 mpy) from pH of 10 to pH of 4. Then it shoots up at pH of 3 and becomes catastrophic at 2.5. Also, raising the pH above 10 causes the corrosion rate to fall to a minimum (3 mpy at pH = 12.5) and then starts to rise as the pH increases over 14, as Pourbaix has shown. This same scientist has created a series of diagrams showing the relationship between potential and pH and deciding whether corrosion or immunity may exist. These data are summarized in Table C-1.

We have also found that minor compositional differences, such as those between cast iron and carbon steel, have no effect on the corrosion resistance.

Table C-1
Corrosion of Steel in Aerated Water
(Based on Uhlig and Others)*

Oxygen Content (ml O_2/1000 ml H_2O)	Corrosion Rate (mpy)	Temperature (°F)	pH
0	0.00	77	7.0
2	4.93	77	7.0
6 (Air saturation)	9.86	77	7.0
10	11.87	77	7.0
13	12.42	77	7.0
15	10.59	77	7.0
17	5.48	77	7.0
20	2.19	77	7.0
25	1.46	77	7.0
6	9.86	77	7.0
6 (Closed system)	20.00	132	7.0
6 (Open system)	18.00	132	7.0
6 (Closed system)	30.00	187	7.0
6 (Open system)	10.00	187	7.0
6	9.86	77	7.0
6	9.86	77	4.0
6	15.00	77	3.0
6	+40.00	77	2.6
6	9.86	77	10.0
6	3.00	77	12.0
6	5.00	77	14.0
6	13.00	77	16.0

* Uhlig, H. H., *Corrosion and Corrosion Control: An Introduction to Corrosion Science and Engineering, 2nd Edition*, 1971, John Wiley and Sons, Inc., New York, pp. 94–95, 99.

We have mentioned steel in water, and this may well describe steel in pipe lines offshore or in rivers. But the question arises: How about steel in the soil?

Actually, steel in deaerated and dry soil should not corrode at all and does not when anaerobic bacteria are absent. But most soils are not dry. The soil resistivities are an indication that moisture and dissolved salts are present, and the corrosivity of the soil is almost proportional to the decrease in resistivity.

The following is a summary of Table C-2 and shows the relation of soil resistivity to corrosion rate of steel in soils based on 12-year tests of

Table C-2
Corrosion of Steel in Soil
(Bureau of Standards Tests and Others)

	Corrosion (mpy)	Type	Soil Resistivities (Ω/cm)
Average of 44 soils	61	Moderately corrosive	1000 to 2000
Tidal marsh	100	Corrosive	500 to 1000
California clay	137	Very corrosive	Below 500
Sandy loam (New England)	21	Mildly corrosive	2000 to 10,000
Desert sand (Arizona)	5	Noncorrosive	Above 10,000

From M. Romanoff, *Underground Corrosion*, Circ. 579, National Bureau of Standards (U.S.) 1957.

buried specimens by the U.S. Bureau of Standards. A soil is considered "noncorrosive" if the soil resistivity is above 10,000 ohm-cm. Between 2000 and 1000 ohm-cm, it is considered "mildly corrosive." Between 500 and 1000 ohm-cm, put it in the "corrosive" class. Below 500 ohm-cm is a special situation requiring immediate action, since an average bare pipe line will corrode in less than a year. This is the "very corrosive" class.

Also, as we have indicated in Chapter 2, sometimes it is necessary to convert readings from other reference electrodes to the Cu-CuSO$_4$ electrode. This may be done using Table C-3, taken from A. W. Peabody's chapter on cathodic protection in the *NACE Basic Corrosion Course*.

Table C-4 is the electromotive series of metals. For each listed metal, the potential given is the *standard electrode potential*. This is determined by placing an electrode of the pure metal in a "standard" solution of its own ions and measuring the potential difference between it and a standard hydrogen electrode—to which is assigned the arbitrary value of zero. The standard solution adopted is that which contains an ion concentration of one mole per 1000 grams of water, and the standard temperature for making the determination is 25°C. Actual potentials developed between pairs of electrodes in various solutions and at various concentrations can vary from the values shown, but the general order is

Table C-3
Comparison of Other Reference Electrode Potentials with that of the Copper-Copper Sulfate Reference Electrode at 25°C

Type of comparative reference electrode	Structure-to-comparative reference electrode reading equivalent to −0.85 volt with respect to copper sulfate reference electrode	To correct readings between structure and comparative reference electrode to equivalent readings with respect to copper sulfate reference electrode
Calomel (saturated)	−0.776 volt	Add −0.074 volt
Silver-Silver Chloride (0.1 N KCl Solution)	−0.822	Add −0.028
Silver-Silver Chloride (Silver screen with deposited silver chloride)	−0.78	Add −0.07
Pure Zinc (Special high grade)	+0.25*	Add −1.10

* Based on zinc having an open circuit potential of −1.10 volt with respect to copper sulfate reference electrode. (From A. W. Peabody, Chapter 5, p. 59, *NACE Basic Corrosion Course*, Houston, TX, 1973.)

the same in most situations; i.e., no metal moves very far from the position shown.

Table C-5 presents the electrochemical equivalents of the metals. This is the amount of the metal which is plated out at a cathode, or dissolved from an anode, expressed as a function of current and time. The values shown are those corresponding to 100% electrochemical efficiency; in actual practice, efficiencies obtained may vary from zero (e.g., shelf deterioration of a dry cell from which no current is being drawn) to very close to 100% (silver coulometer operated under careful laboratory conditions). The efficiency of magnesium anodes in ordinary applications generally runs about 50%; it may go as high as 75% at high current densities under favorable conditions. The efficiency of zinc anodes is in many cases higher, sometimes reaching 95%.

Table C-4
Electromotive Series of Metals

Metal	Ion Formed	Potential
Lithium	Li^+	+2.96
Rubidium	Rb^+	+2.93
Potassium	K^+	+2.92
Strontium	Sr^{++}	+2.92
Barium	Ba^{++}	+2.90
Calcium	CA^{++}	+2.87
Sodium	Na^+	+2.71
Magnesium	Mg^{++}	+2.40
Aluminum	Al^{+++}	+1.70
Beryllium	Be^{++}	+1.69
Manganese	Mn^{++}	+1.10
Zinc	Zn^{++}	+0.76
Chromium	Cr^{++}	+0.56
Iron (ferrous)	Fe^{++}	+0.44
Cadmium	Cd^{++}	+0.40
Indium	In^{+++}	+0.34
Thallium	Tl^+	+0.33
Cobalt	Co^{++}	+0.28
Nickel	Ni^{++}	+0.23
Tin	Sn^{++}	+0.14
Lead	Pb^{++}	+0.12
Iron (ferric)	Fe^{+++}	+0.04
Hydrogen	H^+	0.00
Antimony	Sb^{+++}	−0.10
Bismuth	Bi^{+++}	−0.23
Arsenic	As^{+++}	−0.30
Copper (cupric)	Cu^{++}	−0.34
Copper (cuprous)	Cu^+	−0.47
Tellurium	Te^{++++}	−0.56
Silver	Ag^+	−0.80
Mercury	Hg^{++}	−0.80
Palladium	Pd^{++}	−0.82
Platinum	Pt^{++++}	−0.86
Gold (auric)	Au^{+++}	−1.36
Gold (aurous)	Au^+	−1.50

Table C-5
Electrochemical Equivalents of Metals

Metal	gm/amp-hr	lb/amp-yr
Lithium	0.259	5.00
Rubidium	3.189	61.63
Potassium	1.458	28.18
Strontium	1.635	31.59
Barium	2.562	49.52
Calcium	0.748	14.45
Sodium	0.858	16.58
MAGNESIUM	0.454	8.77
ALUMINUM	0.335	6.48
Beryllium	0.168	3.25
Manganese	1.025	19.98
ZINC	1.220	23.57
Chromium	0.647	12.50
IRON (ferrous)	1.042	20.14
Cadmium	2.097	40.52
Indium	1.427	27.58
Thallium	2.542	49.00
Cobalt	1.099	21.25
Nickel	1.095	21.15
Tin	2.214	42.80
Lead	3.865	74.70
IRON (ferric)	0.695	13.42
HYDROGEN	0.038	0.72
Antimony	1.514	29.26
Bismuth	2.599	50.14
Arsenic	0.932	18.00
Copper (cupric)	1.186	22.92
Copper (cuprous)	2.372	45.83
Tellurium	1.190	23.00
Silver	4.025	77.78
Mercury	3.742	72.32
Palladium	1.990	38.46
Platinum	1.821	35.19
Gold (auric)	2.452	47.39
Gold (aurous)	7.357	142.18

Table C-6 shows the heat of formation of the chlorides of the various metals. Values shown are for the combination of a single chloride ion; that is, the total heat of formation has in each case been divided by the valence of the metal. It will be observed that this ranking, by chemical affinity or activity, is very similar in order to that in Table C-1, where the order is determined by electrical activity.

Table C-6
Heat of Formation Chlorides of Metals
(Values for Single Ion)

Metal	Heat of Formation	Metal	Heat of Formation
Rubidium	105.0	Tin	40.6
Potassium	104.3	Cobalt	38.5
Barium	102.7	Nickel	37.5
Strontium	98.9	Copper (cuprous)	32.5
Sodium	98.4	Iron (ferric)	32.1
Lithium	97.4	Silver	30.6
Calcium	95.4	Antimony	30.5
Magnesium	76.6	Bismuth	30.2
Manganese	56.3	Mercury	26.7
Beryllium	56.2	Copper (cupric)	25.7
Aluminum	55.6	Arsenic	24.1
Zinc	49.8	HYDROGEN	22.0
Chromium	49.8	Palladium	21.7
Thallium	48.7	Tellurium	19.3
Cadmium	46.5	Platinum	15.6
Indium	42.9	Gold (aurous)	10.3
Lead	42.8	Gold (auric)	9.0
Iron (ferrous)	40.9		

Appendix D
Attenuation Equations

In Chapter 5, Figure 5-2 shows the attenuation curves for an infinite line. The equations for these curves are derived from the following assumptions:

1. The leakage conductance to earth is uniform for the entire line; this requires that the coating conductance and the soil resistivity be uniform.
2. Current is drained from the line at a single point and discharged to earth at a remote distance so that no trace of the anode field is detectable at the pipe line.
3. The resistance of the pipe per unit length is uniform throughout.
4. The line is infinitely long.

Under these conditions, the equations are:

$$\Delta E_x = \Delta E_o e - \alpha x \tag{D-1}$$

$$\Delta I_x = \Delta I_o e - \alpha x \tag{D-2}$$

where:
ΔE_X = change in pipe-to-soil potential at distance x from the drainage point

ΔE_O = change in the same quantity at the drainage point
e = base of natural logarithms = 2.718 . . .
α = quantity called the attenuation constant
α is a function of the particular line involved and is given by

$$\alpha = R_s/R_k \qquad \text{(D-3)}$$

where R_s is the longitudinal resistance of the pipe in ohms per foot, and R_k is the "characteristic resistance" of the line, in ohms. This in turn is given by

$$R_k = \sqrt{R_s R_1} \qquad \text{(D-4)}$$

where R_1 is the leakage resistance to remote earth of the line in ohm-feet.

The characteristic resistance is the resistance of the whole line to remote earth, as seen from the drainage point, looking in one direction only; it is the ratio $\Delta E_o/\Delta I_o$. It will be noted that the unit for α is "per foot."

Equations D-1 and D-2 exhibit the attenuation of potential and voltage along the line; being exponential in form, they will plot as straight lines on semilog paper (see Figure 5-2).

Consider now a section of line of length $2L$ lying between two identical drainage points; let the remainder of the original assumptions made still hold. Under these conditions, at any point at a distance x from one of the two points (chosen as an origin) the potential will be given by the sum of two expressions similar to Equation D-1, and the current by the difference of two similar to Equation D-2:

$$\Delta E_x = \Delta E_o' e - \alpha x + \Delta E_o' e \ (2L\text{-}X) \qquad \text{(D-5)}$$

$$\Delta I_x = \Delta I_o' e - \alpha x - \Delta I_o' e - \alpha \ (2L\text{-}X) \qquad \text{(D-6)}$$

where $\Delta E_o'$ and $\Delta I_o'$ are the potential shift and line current which would be caused by either of the two units, at its drainage point, if the other unit were not present. The actual ΔE_o and ΔI_o, it can be seen, will be somewhat different. From these two, the following equations can be derived:

$$\Delta E_x = \Delta E_o \cosh \alpha \ (L\text{-}x)/\cosh \alpha \ L \tag{D-7}$$

$$\Delta I_x = \Delta I_o \sinh \alpha \ (L\text{-}x)/\sinh \alpha \ L \tag{D-8}$$

where ΔE_o and ΔI_o now refer to the actual drainage potential and current (note that ΔI_o is always the current drained from one direction only and is normally half the total drainage current).

These two curves are shown in Figure 5-4. They were derived as curves showing conditions along half of a line segment between two identical drainage points; they may be used, however, to represent conditions along a line between a drainage point and an insulated joint lying at distance L; for the insertion of an insulated joint at the point of zero current, there will be no effect on the line whatever. These are the curves represented in Figure 5-3.

It may be argued that the assumptions of uniformity made in the derivation of the attenuation equations are so severe that no actual line approaches the conditions closely enough for the equations to be useful. This is not the case, however; they are quite useful, but the assumptions must always be remembered, and it is true that many actual lines deviate so far from uniformity that they are not applicable. Equations D-5 and D-6 will be found to conform to actual field conditions more often than the "infinite" line Equations D-1 and D-2; these, however, are frequently applicable to bare or poorly coated lines. It should also be noted that the better the coating on a line, the less is the influence of soil resistivity on total leakage conductance; a well-coated line will often show very uniform attenuation characteristics, although it passes through soil which is far from uniform.

It is always helpful to plot the values of ΔE and ΔI from a current requirement test on semilog paper and give them a critical look. Knowing what the curves should look like for the ideally uniform conditions will often make it possible to determine the causes of the anomalies present in the data, with or without the actual use of mathematical analysis.

Index

Printed and bound by CPI Group (UK) Ltd, Croydon, CR0 4YY

03/10/2024

01040432-0003